Lecture Notes in Computer Science 6555

Commenced Publication in 1973
Founding and Former Series Editors:
Gerhard Goos, Juris Hartmanis, and Jan van Leeuwen

Jan Camenisch   Valentin Kisimov
Maria Dubovitskaya (Eds.)

Open
Research Problems
in Network Security

IFIP WG 11.4 International Workshop, iNetSec 2010
Sofia, Bulgaria, March 5-6, 2010
Revised Selected Papers

 Springer

Volume Editors

Jan Camenisch
Maria Dubovitskaya
IBM Research Zurich, Säumerstr. 4
8803 Rüschlikon, Switzerland
E-mail: {jca, mdu}@zurich.ibm.com

Valentin Kisimov
University of National and World Economy
Studentski Grad "Hr. Botev", 1700 Sofia, Bulgaria
E-mail: vkisimov@gmail.com

ISSN 0302-9743   e-ISSN 1611-3349
ISBN 978-3-642-19227-2   e-ISBN 978-3-642-19228-9
DOI 10.1007/978-3-642-19228-9
Springer Heidelberg Dordrecht London New York

Library of Congress Control Number: 2011920837

CR Subject Classification (1998): K.6.5, K.4, C.2, E.3, D.4.6, H.3.4-5

LNCS Sublibrary: SL 4 – Security and Cryptology

Typesetting: Camera-ready by author, data conversion by Scientific Publishing Services, Chennai, India

Printed on acid-free paper

Springer is part of Springer Science+Business Media (www.springer.com)

Preface

iNetSec 2010 is the main conference of working group WG 11.4 of IFIP. Originally, the conference was run in the traditional format where research papers get submitted, peer-reviewed, and then presented at the conference. Because there are (far too) many security conferences like that, it was decided in 2009 to change the format into a forum for the discussion of open research problems and directions in network security.

To enable this more open style, while still remaining focused on particular topics, we called for two-page abstracts in which the authors were asked to outline an open research problem or direction. These abstracts were reviewed by the entire program committee who ranked each of them according to whether the problem presented was relevant and suited for a discussion. Based on this, about half of the submitted abstracts were chosen for presentation and discussion at the conference. The authors were asked to later write and submit full papers based on their abstracts and the discussions at the workshop. These are the papers that you are now holding in your hands.

The conference also hosted two invited talks. Basie von Solms (President of IFIP) argued in his talk entitled "Securing the Internet: Fact or Fiction?" that secure computer networks are an illusion with which we have to cope. The paper to the talk is also contained in these proceedings. Leon Straus (President Elect of IFIP) shared his insights on "Network and Infrastructure Research Needs from a Financial Business Perspective" which showed how much research can and should learn from practitioners from all fields using computer technologies.

On the last day of the conference, the attendees gathered in a lively discussion about security and cloud computing that opened the eyes of quite a few. The social highlights of the conference were the Bulgarian dinner accompanied by traditional live music and a guided tour through Sofia that despite the freezing temperature was delightful and impressive.

We are grateful to the two invited speakers, the authors, the PC members, and last but certainly not least, the local organizing committee.

September 2010

Jan Camenisch
Valentin Kisimov

iNetSec 2010

Open Research Problems in Network Security

University of National and World Economy, Sofia, Bulgaria
March 5-6, 2010

Organized in cooperation with *IFIP WG 11.4*

Executive Committee

Program Chair	Jan Camenisch, IBM Research – Zurich
Organizing Chair	Valentin Kisimov, UNWE

Program Committee

Jan Camenisch	IBM Research
Virgil Gligor	Carnegie Mellon University
Jean-Pierre Hubaux	EPFL
Simone Fischer-Hübner	Karlstad University
Dogan Kesdogan	University of Siegen
Valentin Kisimov	UNWE
Albert Levi	Sabanci University
Javier Lopez	University of Malaga
Refik Molva	Eurecom

Local Organizing Committee

Valentin Kisimov	IFIP TC11, UNWE
Dimiter Velev	UNWE
Kamelia Stefanova	UNWE
Vanya Lazarova	UNWE

Table of Contents

Invited Talk and Scheduling

Securing the Internet: Fact or Fiction?................................. 1
    Basie von Solms

Open Research Questions of Privacy-Enhanced Event Scheduling........ 9
    Benjamin Kellermann

Adversaries

Event Handoff Unobservability in WSN 20
    Stefano Ortolani, Mauro Conti, Bruno Crispo, and Roberto Di Pietro

Emerging and Future Cyber Threats to Critical Systems 29
    Edita Djambazova, Magnus Almgren, Kiril Dimitrov, and
    Erland Jonsson

Adversarial Security: Getting to the Root of the Problem 47
    Raphael C.-W. Phan, John N. Whitley, and David J. Parish

Practical Experiences with Purenet, a Self-learning Malware Prevention
System 56
    Alapan Arnab, Tobias Martin, and Andrew Hutchison

A Biometrics-Based Solution to Combat SIM Swap Fraud 70
    Louis Jordaan and Basie von Solms

Are BGP Routers Open to Attack? An Experiment 88
    Ludovico Cavedon, Christopher Kruegel, and Giovanni Vigna

Secure Processes

Securing the Core University Business Processes 104
    Veliko Ivanov, Monika Tzaneva, Alexandra Murdjeva, and
    Valentin Kisimov

Some Technologies for Information Security Protection in Weak-
Controlled Computer Systems and Their Applicability for eGovernment
Services Users 117
    Anton Palazov

Real-Time System for Assessing the Information Security of Computer
Networks   123
 Dimitrina Polimirova and Eugene Nickolov

Evidential Notions of Defensibility and Admissibility with Property
Preservation.   134
 Raphael C.-W. Phan, Ahmad R. Amran, John N. Whitley, and
 David J. Parish

Security for Clouds

Cloud Infrastructure Security.   140
 Dimiter Velev and Plamena Zlateva

Security and Privacy Implications of Cloud Computing – Lost in the
Cloud   149
 Vassilka Tchifilionova

The Need for Interoperable Reputation Systems   159
 Sandra Steinbrecher

Author Index.   171

Securing the Internet: Fact or Fiction?

Basie von Solms

President: IFIP
University of Johannesburg
Johannesburg, South Africa
`basievs@uj.ac.za`

1   Introduction

The number of users of the Internet, in whatever way, is growing at an explosive rate. More and more companies are rolling out new applications based on the Internet, forcing more and more users to leverage these systems and therefore become Internet users. Social networking sites and applications are also growing at alarming rates, getting more and more users, whom we can call home or private users, involved and active on the Internet. Corporate companies are now also integrating social networking as part of their way of doing business, and governments are implementing Internet based systems ranging from medical applications to critical IT infrastructure protection.

Therefore millions of people are using the Internet for e-commerce, information retrieval, research, casual surfing and many other purposes, and this will just keep growing. One estimate is that the amount of data space needed to support the fast growing online economy will double every 11 hours by 2012 [1].

In most, if not all, of these activities on the Internet, data and information are involved. This may range from the user's IP address to secure personal information, to sensitive corporate information to crucial national strategic information.

The big question is, and has always been, how secure is all this information and data, and can it be properly secured?

This of course has worried experts even since information was stored in electronic systems many years ago, but has become much more acute the last few years with the explosive growth of the Internet, and the, in many cases, uncontrolled flocking of users to be part of Internet usage in some way or the other.

This paper objectively investigates this question, based on recent reports in the area of Information Security and Cyber Crime.

In paragraph 2 we will investigate what can we understand under the term "securing" of the Internet - what do we want to secure and why. In paragraph 3 we will give an overview of recent cyber crime statistics, and in Section 4 we will give an opinion about the possibility of securing the Internet.

2   What Do We Mean by "Securing" the Internet?

It is important to investigate what we have in mind when we say we want to secure the Internet, as there surely are different interpretations of this term. In this paragraph we will look at possible meanings of the term.

J. Camenisch, V. Kisimov, and M. Dubovitskaya (Eds.): iNetSec 2010, LNCS 6555, pp. 1–8, 2011.

2.1   The Ideal Interpretation

In the ideal and widest possible interpretation, we can say that securing the Internet means that all data and information stored on all websites forming part of the Internet, and all data and information being transported over the Internet are secured so that no unauthorized people can see (read) or change the content (protecting the confidentiality and integrity of the data and information), and that the data and information must be available to authorized users whenever they want to use it (protecting the availability of the data and information.)

This means that all information in all databases and in transit must be secured and only accessible to the authorized users. Apart from extensive encryption techniques to ensure this, extensive identification and authentication techniques must be available to ensure that every user is correctly identified and authenticated, and logical access control is comprehensively enforced. These measures must ensure that no person can masquerade as another person, and that electronic identities can only be successfully used by the real owners to which such identities were issued.

To create this ideal interpretation, the following must be possible:

- we must know precisely what is part of the Internet, ie which computers, servers and other equipment
- all these infrastructure must be controlled via legal systems to enforce the required confidentiality, integrity and availability
- no unauthorized system can be connected to the Internet - authorized by some (central?) managing power
- all users' identity information must be protected in such a way that it can never be compromised in any way
- all users must be absolutely aware of the risks of compromising their identity information
- legal systems must exist internationally which can enforce these requirements
- etc etc

Even a person with no information security knowledge at all, will agree straight forward that this ideal situation is NOT possible, because of the open and uncontrolled way the Internet is operated and growing.

Therefore, in the light of the ideal interpretation of securing the Internet, as discussed above, we must conclude

Securing the Internet is fiction - it is just not possible.

2.2   The Realistic Interpretation

Let us now investigate a more realistic interpretation. In this realistic interpretation, we have to accept that

- there is no, and never will be any, central control over the Internet
- there is no way to know what the boundaries of the Internet are, i.e. which systems are part of the Internet at any point of time

- no legal systems exist (presently) which can enforce any reasonable security (like enforcing encryption and proper identification and authentication)
- users do not protect their identity information, and are in most cases not aware of the risk of not doing so
- in no way can masquerading, or unauthorized use of identity information be prevented, as users are the weakest link in the chain and can always be seduced to compromise their identity information
- cyber crime is rampant and leveraging any possible chink in the armor of the Internet

Therefore, in the light of this realistic interpretation of securing the Internet as discussed above, the author concludes

Securing the Internet is fiction - it is just not possible.

This conclusion will be motivated in the next paragraphs.

3   An Overview of Recent Cyber Crime Statistics

An overview of recent cyber crime statistics, provides a good place on which to start and base a motivation for the view taken in the previous paragraph. In this paragraph, we will investigate some recent international reports on cyber crime, and try to get an impression of what is happening. The paragraph will be unstructured in the sense that we will provide a few quotes from specific reports, and then briefly comment on these in paragraph 4. Paragraph 5 will provide some suggestions for the future.

3.1   The Sophos Security Threat Report - 2009 ([2])

Under the summary, Six Months at a glance:

- 23 500 infected websites are discovered every day. That's one every 3.6 seconds - four times worse than the same period in 2008
- 15 new bogus anti-virus vendor websites are discovered every day. This number has tripled, up from average of 5 during 2008
- 89.7% of all business email is spam

The report further makes the following very worrying statement:

"The vast majority of infected websites are in fact legitimate sites that have been hacked to carry malicious code. Users visiting the websites may be infected by simply visiting affected websites, ... The scope of these attacks cannot be underestimated, since all types of sites - from government departments and educational establishments to embassies and political parties ... - have been targeted."

3.2   The CISCO White Paper ([3])

The White paper states that *"Internet users are under attack. Organized criminals methodically and invisibly exploit vulnerabilities in websites and browsers and infect computers, stealing valuable information (login credentials, credit card numbers and intellectual property) and turning both corporate and consumer networks into unwilling participants in propagating spam and malware"*.

An extremely worrying aspect reported is the fact that *"trusted legitimate websites are the perfect vehicle for malware distribution. (it is estimated) that more than 79% of the websites hosting malicious code are legitimate websites that have been exploited"*.

3.3   The UK Cybercrime Report 2009 ([4])

The report indicate that during 2008, *"cyber criminals committed over 3.6 million criminal acts online (that is one every 10 seconds)"*. Furthermore, *"online banking fraud has increased by a staggering 132%, with losses of UKP 52.5 million, compared to UKP 22.6 million in the previous year. This sharp rise can mostly be attributed to nearly 44 000 phising sites specifically targeting banks and building societies in the UK. Phising sites are becoming more prevalent and increasingly sophisticated."*

3.4   CISCO Annual Security Report 2009 ([5])

In this report, the following aspects are mentioned, which can let Internet users sleep uneasy at night!

"According to the Anti-Phising Working Group, the number of fake anti-virus programs grew by 585% from January to June 2009. Banking Trojans, like Zeus and Clampi, increased by nearly 200%".

An extremely worrying aspect highlighted in the report, is the growing use of mobile phones and text messages to lure the user into visiting infected websites. The report warns that this will have a huge impact on users as the use of mobile phones grows to gain access to the Internet.

"Expect to see more "smishing" scams (phishing attacks using SMSs) in 2010" warns the report.

Furthermore *"as more individuals worldwide gain Internet access through mobile phones (because in many parts of the world it's faster to than waiting on available broadband), expect cyber crime techniques that have gone out of fashion to re-emerge in many developing countries. Cyber criminals will have millions of inexperienced users to dupe with unsophisticated or well-worn scamming techniques that more savvy users grew wise to (or fell victim to) ages ago."*

The report specifically refers to the use of other ways and approaches being used more and more for formal company work.

"Where does work happen? No longer does business take place solely behind network walls. The critical work of an organization is happening increasingly on social networks, on handheld devices, on Internet kiosks at airports and at local cafes. (This will) *demand a new way of thinking about how to secure an enterprise ... "*

3.5   T3.com ([6])

"There is now an epidemic of Malware online. Some experts are saying that there will come a point very soon where you can barely open a web page without coming under attack. ... anything up to half a million computers could be falling victim to "drive-by downloads" each day. If you surf without any electronic safeguard, the chances are that you are already one of them".

In the same article, Mikko Hypponen from F-Secure is quoted as saying *"However, even though we are not winning, we are certainly not giving up either."*

3.6   The Washington Post ([7])

"Law enforcement agencies worldwide are losing the battle against cyber crime. (the) number of compromised PCs used for blasting out spam and facilitating a host of online scams has quadrupled in the last quarter of 2008 alone, creating armies of spam "zombies" capable of flooding the Internet with more than 100 billion spam messages daily"

After this (sobering?) overview, let us now evaluate the situation in the next paragraph, returning to the author's view stated in paragraph 2.

4   Securing the Internet Is Fiction – It Is Just Not Possible

In this paragraph we will try to evaluate some of the aspects mentioned in the previous paragraph, and try to determine whether such aspects can be addressed in a way which can help to properly secure the Internet.

4.1   The Malicious Software (Malware) and Spoofing Threats

Can this threat be properly addressed? With legitimate web sites, for eg those of your favorite and trusted search engine, bank, sports club etc now being infected by malware, and your computer becoming infected by just visiting such a website, how will you know you are infected? Even if you have the latest anti-virus (AV) software, the malware may not be recognized by the AV software, or the AV software may in itself be infected.

To a certain extent it can be expected that large corporate enterprises, having large security budgets, and a brand name to protect, may have advanced methods to detect such malware and protect their IT infrastructure, but it may not help! The newest type of malware may not be recognized by AV software, no matter how up to date they are. Such corporations still have to accept that their employees, logging onto the company IT infrastructure from unsafe sites, may cause infection - even if the strictest policies and procedures are in place.

However, the bigger worry should be about the home user, and even the small and medium enterprise who cannot afford to spend large amounts (even small amounts for that matter) on security of their systems, either because they cannot afford it, or they are not aware of the risks!!

How is this home user or small company user going to recognize a spoofed website? Even the security savvy experts have fallen for sites which are so well spoofed, that they just cannot be recognized as such? With 44 000 sites just concentrating their efforts on spoofing banking sites in the UK (see 3.3 above), the expertise and knowledge of those creating and operating these sites, must be such that the ordinary unsophisticated user will not recognize it!

Are users in developing countries going to be first put through awareness courses to sensitize them about potential risks and scams before they are granted mobile based Internet banking access, or are they just going to be left out there to fight for themselves (and lose their money?) (see 3.4 above).

It must take a very optimistic person (maybe with blinkers on), to state that all such users can, and will, be secure in venturing on the Internet.

The author's conclusion:

Trying to totally protect users in this environment is fiction.

4.2   The Threat of an Insufficient Legal Infrastructure for Curtailing Cyber Crime

Can we create and agree on a legal environment, accepted by all countries in the world, to ensure that all cyber criminals are caught and brought to book?

I am sure that even our very optimistic person of the previous paragraph, will agree that this is not possible!!

The author's conclusion:

To hope for a comprehensive, internationally agreed and cross border enforceable set of laws and penalties, to curtail cyber crime, is fiction.

4.3   The Threat of Error Prone Software Being Rolled Out

Many sites are infected by malware because of some error or flaw in the system software supporting user transactions - from the browser software through to the operating system software and other utilities. Such errors and flaws arise because of the immense complexity of these software systems - consisting of millions lines of code. These errors and flaws are then exploited by malware writers to infect such systems. Patches are provided by the original suppliers when these errors become known, but some systems are never patched by their users, some errors are exploited before they can be patched, and some errors are exploited by the discoverer, and never reported!

Can such systems be developed, with the present level of software engineering knowledge, without such inherent flaws and errors? Can such systems always be patched in time before exploitation? Will all errors be discovered before exploitation?

The author's conclusion:

To hope for perfect software without flaws and errors is fiction.

5   Final Conclusion

From the discussion above, the author's conclusion is that presently, and with the present knowledge:

Securing the Internet is fiction.

6   So What About the Future?

Being negative about comprehensively securing the Internet and the users using the Internet, very definitely does not mean that the author is of the opinion that users should stop using the Internet - far from it! The positive side (benefits) of using the Internet, at least at this stage, still outweighs the negative side (disadvantages).

The Internet will only grow, and it will just keep serving more and more applications.

Any person who wants to buy a car, must have a driver's license. To obtain such a license, the person must complete a strict training course in which the potential driver is exposed to all the laws, rules, regulations and risks related to driving a car. If the user passes, and buys a car, he/she is doing that on an informed basis about the risks of driving a car, and the possibility of being injured or even killed in an accident. One thing you are never promised, and which every driver understands, is that there are NO promises by anyone that no mishaps will occur when driving. Driving without a license is an offense, in your own country and across borders.

The author's preferred solution lies in massive campaigns and controls to make users absolutely aware of the risks of using the Internet, so that they can take informed decisions whether they want to take the risk - they should know that they must take responsibility for their own security on the Internet, and should never think that someone else will protect them!

What must happen as far as Internet usage is concerned, and which must happen on a massive national and international scale, is that all users must be made aware of the risks of using the Internet. This will probably mean compulsory security awareness courses (with proof of understanding?) before a user is granted Internet access rights. The user should be in a position to make a well informed decision of whether the benefits for which he/she wants to use the Internet, outweighs the risks. One thing users should never be promised, and which every Internet user should understand, is that there are NO promises by anyone that no mishaps will occur when accessing the Internet. Users should understand and accept that accessing the Internet has risks, and that it is not secure or safe.

We need an Information Security Internet Driver's License (ISIDL). Such a license should prepare and inform the user about the risks, related to potential financial fraud, loss of privacy, the protection of identity information etc, in accessing the Internet. Accessing the Internet without an ISIDL should be an offense, in your own country and across borders.

7   The End

- Accessing the Internet has many risks
- Securing the Internet is fiction and impossible
- Allowing users to access the Internet without such users being totally aware of relevant risks, is unethical, and should be prohibited
- Comprehensive national campaigns to inform users of the risks of Internet usage (to be run by Governments, introduced into schools etc)
- Comprehensive User Internet Security Awareness should be the entrance requirement before access is granted.

References

1. ITNow (January 2010), http://www.bcs.org
2. Sophos, The Sophos Security Threat Report (2009),
   http://www.sophos.com/sophos/docs/eng/marketing_material/
   sophos-security-threat-report-jan-2009-na.pdf
3. CISCO, A Comprehensive Proactive Approach to Web based Threats, CISCO Iron-Port Web Reputation White Paper (2009),
   http://www.ironport.com/pdf/ironport_web_reputation_whitepaper.pdf
4. UK Cybercrime, The UK Cybercrime Report (2009),
   http://www.garlik.com/press.php?id=613-GRLK_PRD
5. CISCO Annual Security Report (2009),
   http://www.cisco.com/en/US/prod/collateral/vpndevc/cisco_2009_asr.pdf
6. T3.com (September 2009), http://www.T3.com
7. Washington Post, Cybercrime is winning the battle over Cyberlaw (2008),
   http://voices.washingtonpost.com/securityfix/2008/12/report_
   cybercrime_is_winning_t.html (accessed February 15, 2010)

Open Research Questions of Privacy-Enhanced Event Scheduling

Benjamin Kellermann

Technische Universität Dresden
Institute of Systems Architecture
D-01062 Dresden, Germany
`Benjamin.Kellermann@tu-dresden.de`

Abstract. Event-scheduling applications like Doodle have the problem of privacy relevant information leakage. A simple idea to prevent this would be to use an e-voting scheme instead.

However, this solution is not sufficient as we will show within this paper. Additionally we come up with requirements and several research questions related to privacy-enhanced event scheduling. These address privacy, security as well as usability of privacy-enhanced event scheduling.

Keywords: event scheduling, electronic voting, superposed sending, anonymity, privacy-enhanced application design.

1 Introduction

For years, technical event scheduling has been done with applications like Microsoft Exchange or a central iCal-server within some intranet. Since Web 2.0 applications became very popular, an application named Doodle [1] which assists event scheduling in a very easy way became popular as well. In this solution, an initiator configures an online poll where everybody votes on dates that are shown to all visitors of the web page.

In contrast to the intranet solution, Doodle offers its event scheduling service to everyone. However, both solutions – the intranet solution and Doodle – have in common that everybody has to publish his so-called *availability pattern*. In contrast to a solution running in a company, within the Doodle-solution, the availability patterns are visible to the whole world.

An availability pattern contains at least two different types of information. The *direct inference* from the pattern are information which one can extract from the availability of certain dates ('Will my husband vote for the date of our wedding anniversary?'). Second, the *indirect inference* of a pattern reveals information, when connecting more than one information source ('The availability pattern of user bunny23 looks suspiciously like the one of my employee John Doe!').

However, Doodle offers users some sort of privacy-enhancing feature. An initiator has the possibility to create "hidden polls" where availability patterns are visible only to him. Together with an SSL based connection, one may achieve

J. Camenisch, V. Kisimov, and M. Dubovitskaya (Eds.): iNetSec 2010, LNCS 6555, pp. 9–19, 2011.
© IFIP International Federation for Information Processing 2011

confidentiality for external attackers. However, the Doodle server and the poll initiator still learns the availability pattern.

Doodle also reacted to their users demand of security. In a normal poll, everybody may change everybody's vote. If one does not want to register to preserve one's unlinkability, Doodle offers the possibility to restrict changes to a vote from the initiator only. However, trust in the poll initiator and the Doodle server is needed in this case as well.

The outline of the paper is as following: We first raise requirements which we claim to be necessary for privacy-enhanced event scheduling in Section 2. Section 3 will give a short literature overview. Afterwards (Section 4) we discuss which questions are not fulfilled yet and should be target of further research.

2   Requirements

This section describes requirements for privacy-enhanced event scheduling. Even if there may be more general requirements applicable to privacy-enhanced event scheduling, we want to concentrate on the most restrictive ones which still fit the most use cases (but not all). Stating too general requirements might lead to a scheme, which is unnecessarily bloated and inefficient.

An event scheduling application typically schedules an event in a group of a few dozens of people. Even if there are use cases, where scheduling events in a group of people who do not know each other are imaginable,[1] the most common use case of such an application is to schedule an event within a closed group where most of the participants know each other.

A typical web-based event scheduling application can be divided into 3 phases: poll initialization, vote casting and result publication. Figure 1 depicts these 3 phases. An initiator creates a poll in the poll initialization phase. There he has to define a set of possible time slots when an event might take place. Every user may specify his preferences in the vote casting phase. Finally, in the result publication phase, a time slot is chosen, when the event should take place. We call the rule, which chooses the time slot, the *selection rule*.

In a usable privacy-enhanced event scheduling application everything should be the same as in a normal event scheduling application with the difference that the availability pattern is confidential.

The following requirements should apply to a privacy-enhanced event scheduling application:

Untrusted single entity. As little trust as possible should be placed in any single entity.

Verifiability. Every participant should be able to verify that no other participant has cheated (i. e., every other participant is authorized and behaves according the protocol) and that his preferences has been taken into account.

---

[1] e. g., one needs to schedule a lecture and wants to find out the most acceptable time slot for the students.

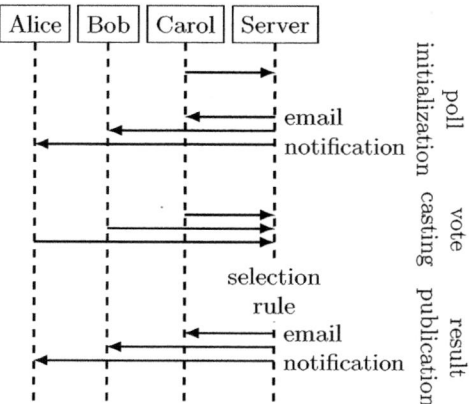

Fig. 1. Phases of a typical web based event scheduling initiated by Carol

Privacy. Nobody should learn more than absolutely necessary about the availability of other participants and thus should not be able to infer on their identity, i. e., every participant should only learn that the one specific chosen time slot fulfills some selection rule. The amount of information which is leaked therefore depends on the selection rule. Note that this requirement does not mean, that all participants are anonymous.

It is easy to conceive, that verifiability and privacy are contradicting requirements. In a Doodle-like scheme without privacy, verifiability is given through the fact, that everybody can read the votes of the other participants. In a privacy-enhanced scheme, other mechanisms are needed.

Additionally, verifiability and privacy require, that the set of participants is fixed within the poll initialization phase and that the participants know each other. Knowing the identity of the other participants gives the possibility to make clear statements about the anonymity set and its size. Additionally one may observe that no attacker is able to add votes for faked participants (only authorized participants vote).

To gain a usable application, the following requirements have to be fulfilled:

Preliminary Steps. An application should not require many preliminary steps (e. g., installation, registration etc.).

Communicational Complexity. An application should not require much more user interaction than existing event schedulers and should allow the user to be offline as it may be used in mobile scenarios. Therefore the scheme has to have few message exchanges. In Figure 1 one can see that typically every user has to interact 2 times with the application: in the vote casting phase and in the result publication. Both interactions are initialized through an email notification.

Computational Complexity. As users do not want to wait long, an application should be efficient for large scheduling problems with many possible event dates.

3 Related Work

Privacy-enhanced event scheduling can be seen as a distributed constraint satisfaction/optimization problem (DSCP/DCOP) or a an e-voting instantiation. We will give a short literature overview of both and explain why they do not fit into the problem. Specific literature is discussed afterwards.

3.1 DSCP/DCOP

A constraint optimization problem consists of a set of variables in finite domains and a set of cost functions. The goal of the optimization is to find a binding for all variables, for which the global cost, calculated from all cost functions is minimal. A constraint satisfaction problem is a special case of the optimization problem where the output of all cost functions is an element of $\{0, 1\}$ and the global cost is calculated with the multiplication of all cost functions. In a distributed constraint optimization, every participant holds his own set of variables and cost functions. The solution to the optimization problem is found through sending messages between the participants with the assignment of variables and their costs.

There exist many algorithms for DCSP [2–4] and DCOP [5–7] and measurements of the information leakage were done by Franzin [8] and Greenstadt [9]. These algorithms may solve complex scheduling problems where different subsets of the participants participate in different events. Each participant may have cost functions about the place, travel time, and constraints between the events.

However, all DCOP algorithms share the problem that they are complex in terms of message exchanges even for basic scenarios. To solve the problem of message exchanges, agents are used which send and receive the messages. As usual users do not want to setup such an agent at some server, they have to run it locally and therefore have to be online at the same time.

Therefore the DCOP approach is too complex in terms of message exchanges and contradicts the communicational complexity requirement. A simpler solution for the simpler problem of scheduling a single meeting would be appropriate.

3.2 E-Voting

There is a lot of literature about electronic voting. It can be categorized into approaches based on mixes [10–14], homomorphic encryption [15–18], and blind signatures [19–23].

The difference between privacy-enhanced event scheduling and e-voting, and why e-voting cannot be applied directly to event scheduling was already discussed [24]. One of the main design criteria of electronic voting schemes is to have a computation and communication complexity, which is independent from the number of participants (voters). While this is valid when organizing an election for millions of citizens, this design criterion can be relaxed in the event scheduling approach, as we deal with smaller closed groups of participants. A design criterion for event scheduling should be that the computational complexity of the scheme scales in the number of time slots. However, if one applies e-voting schemes directly to event scheduling, they all have in common that the number of asymmetric operations scales linear with the number of time slots, i. e., one needs at minimum one asymmetric operation per time slot.

In the following, we will shortly discuss the three approaches separately and estimate the computational complexity in terms of asymmetric operations. Let $|T|$ be the number of possible time slots an event may be scheduled at, and $|P|$ be the number of participants voting in a poll.

Mixes. Chaum invented mixes as building blocks for anonymous communication channels and he first proposed an election scheme based on them [10]. Many extensions have followed thereafter [11–14]. The common idea is to build an anonymous blackboard with mixes. In a first phase, every voter has to generate an asymmetric key pair and publish the public key on that blackboard. In the voting phase, every voter encrypts his or her vote with the secret key and publishes the encrypted vote on the anonymous blackboard. Everybody can decrypt the votes and count the result.

Assuming ℓ mixes, to cast a vote, every voter has to encrypt his or her key and vote ℓ times asymmetrically. A naive adaption to event scheduling would imply one poll per time slot. Every voter would have to do $2 \cdot \ell \cdot |T|$ asymmetric encryptions. Further, one must trust that the mixes do not collude to compromise one's privacy, and the mixes have to perform additional decryption operations, which add to the overall complexity.

Homomorphic Encryption. Voting schemes based on a homomorphic encryption function use the property that one can add all the votes and decrypt the result without decrypting individual votes (i. e., one can find two operations $\oplus$ and $\otimes$ so that the encryption function E is homomorph to these operations $E(x_1) \oplus E(x_2) = E(x_1 \otimes x_2)$). A problem of plain homomorphic encryption is that cheating voters stay undetected as their votes are not decrypted separately. So practical schemes require extra effort to prevent this.

A voting scheme based on a homomorphic encryption function was first described by Cohen and Fischer [15] and later extended by Benaloh and Yung [16]. The scheme consists of different parts where a central entity and the voters have to commit to values and prove their correctness without revealing them. Proving a "primary" ballot is done by committing to several "auxiliary" ones, decrypting half of them and showing that the others are type-equivalent to the primary ballot. This proof is done twice in the whole scheme.

A direct application of this method for event scheduling appears to be inefficient: considering only the vote encryptions, with a security parameter ℓ, every voter would have to do $\ell \cdot |T|$ asymmetric cryptographic operations.

Inspired by Benaloh, Sako and Kilian introduced a voting scheme that uses partially compatible homomorphisms [17]. Baudron et al. enhanced this scheme for multi-votes [18]. However, also in this scheme, a voter has to encrypt every single vote multiple times. Baudron's extension even targets multi-candidate elections, but this is not equivalent to the event scheduling problem, where repeated elections would be needed.

Blind Signatures. Shortly after his mix-approach, Chaum came up with another voting scheme [19] which uses blind signatures [25]. Fujioka et al. reduced the complexity of Chaums idea to adapt it to large scale elections [20] and many subsequent schemes were derived from the Fujioka–Okamoto–Ohta scheme [21–23]. Some of them led to implementations [26, 27].

The main idea is to split the protocol into two independent phases: (1) administration, which handles access control, and (2) counting of anonymously casted votes. The vote is blindly signed during administration, unblinded by the voter and then sent anonymously to the counter. As the casted votes contain no personal information, they can be published afterwards.

If one applies this scheme to the event scheduling problem, a voter would have to blind and unblind $|T|$ messages and verify the administrator's signature of every message. The administrator would have to verify $|P|$ signatures and has to sign $|T|$ messages. The counter would have to verify $|P| \cdot |T|$ signatures. The overall effort is $(|P| + 2) \cdot |T| + |P|$ asymmetric cryptographic operations.

3.3   Specific Prior Work

There are two publications which specifically target the event-scheduling problem [24, 28]. Herlea et al. [28], covers three different approaches to the problem. One is a solution based on a trusted third party, which is efficient, but stands in contrast to the requirement of limited trust in a single entity. The second is a straight application of general secure distributed computing. Consequently, it suffers from high computational and communication complexity. The third approach, a "custom-made negotiation protocol" is most interesting and promising. It can be best described as a hybrid between the techniques reviewed in Section 3. The protocol is designed to schedule single events through a combination of *homomorphic encryption* with respect to the equality operation (in fact, addition modulo two) to *blind* individual availability pattern, and an anonymous channel, which is established by letting voters act as re-encrypting *mixes*. While the cryptographic operations are comparably efficient, the scheme requires way too much communication phases ($|P| + 1$ messages). (The authors acknowledge this and discuss ways to trade off communication complexity against trust assumptions.)

Based on superposed sending and Diffie-Hellman key agreement [29, 30], Kellermann and Böhme described a privacy-enhanced event scheduling solution [24, 31, 32]. Within this solution, every possible participant encrypts his availability pattern with a homomorphic encryption. The availability pattern itself contains a 1 for all available time slots and a 0 for time slots, the participant is unavailable. The homomorphism of the encryption is used afterwards to calculate the number of all available participants at the certain time slots.

4   Open Research Questions

Offering a privacy-enhanced version of a web-based event scheduling system raises several problems which are described in the following and should be target of further research.

4.1   Security Definition and Formal Proof of Correctness

The whole process of multilateral secure voting has assumptions about privacy and security. Security definitions of e-voting or DCOP can already be found in the literature [33]. However, a good application to privacy-enhanced event scheduling as well as a formal proof of correctness is still missing.

A good requirement formalization is still needed. With such a formalization, a proof of correctness of a certain scheme could be done.

4.2   Predefined Complex Selection Rules

A problem a privacy-enhanced event scheduling solution has to deal with is when not all people are available at a certain time slot. The most common selection rule is to choose one of the time slots where most of the people are available. However, it may be desirable to decide on other rules, than solely on the number of available participants. E. g., it may be the case, that some people are more important to be at a meeting than others. Another selection rule may depend on sets of participants, e. g., one may schedule a meeting between a number of organizations where it is necessary that one person of every organization attends the meeting.

When scheduling an event with an existing privacy-invasive solution, one may make such decisions, without the involvement of all possible participants. Even if no common time slot is found where all participants may participate, one is able to select a time slot on other criteria than initially defined. If the selection rule came to no result in a privacy-enhanced scheme one has the problem, that it is not possible to change it without the involvement of all participants. A possible solution may be the specification of more than one selection rule before scheduling an event. The additional selection rules are taken into account when the first one came to no result.

Specifying complex selection rules can be expressed in functions in a DCOP solution. However, e-voting solutions and the specific privacy-enhanced event scheduling approaches suffer these functionality. Here, only the sum of all available participants at the time slots can be taken into account.

A problem is that it is difficult to measure the privacy for an arbitrary selection rule. Measurements which can be found in the DCOP literature may be applied here as well. Furthermore, the amount of privacy has to be displayed to the participants when they state their availabilities. Additionally it is unclear which selection rules with their respecting implications are understandable for average users.

4.3   Specifying Preferences

Within the vote casting phase, every participant has to specify a binary choice if he is available at a certain time slot or not. Instead of sending this binary choice it may be desirable to specify preferences for every time slot. This can already be done in a Doodle poll. There one may create a so called "Yes-No-Ifneedbe"-poll where one has the possibility to select one out of three states for every time slot.

Such a generalization is not necessarily possible in a privacy-enhanced event scheduling solution. E. g., when using homomorphic encryption the votes cannot be taken from an arbitrary domain. Consequently, one may not apply arbitrary operations to the votes. This should be target of further research and one may think of either giving the answers a meaning (like Doodle does) or let the user specify a preference value.

4.4   Prevent "Legal-but-Selfish Votes"

A legal-but-selfish vote is a vote where a participant indicates his availability only at his preferred time slot and not at all time slots he is available. Solutions which only come to a solution if all participants are available at a common time slot (e. g., DCSP) motivate users to anonymously send legal-but-selfish votes.

Within a scheduling solution where all participants can observe all indicated availabilities, the motivation of being selfish would go along with reputation loss. However, in a privacy-friendly solution selfish votes may be sent anonymously.

A possible prevention of selfish votes may let participants prove that they signaled availability for more than a certain minimum number of time slots. Knowing the fact that every voter must be available at a number time slots in order to participate at the poll would of course decrease the privacy of the voters.

4.5   Automatic Poll Termination

It was already mentioned, that it is necessary to fix the set of participants in the poll initialization phase. However, it might occur that a participant should be removed after a poll has started.

Kellermann and Böhme already discussed how to kick-out users who should not participate in the poll anymore. From a usability point of view, this seems to be a valid requirement, as depending on the scheme, the selection rule might be unable to calculate the result, if one of the participants does not vote. If one thinks one step ahead, an open question is how to handle an automatic termination of the poll, either after a specific amount of time, or after a specific number of participants voted. Such an extension should not undermine the privacy of the participants.

4.6   Dynamic Insertion and Deletion of Time Slots

Sometimes it may be necessary to dynamically insert or delete time slots, after the poll has already started. This may occur after participants already casted votes.

Dynamic insertion and deletion of time slots may not be obviously possible in a privacy-enhanced event-scheduling scheme (e. g., it may influence the cost functions of other time slots in a DCOP based solution). If one does, already casted votes should be taken into account. The privacy of votes for all time slots should not be affected, when the set of time slots is changed. Current schemes do not target this issue.

4.7 Updating and Revoking Votes

A feature offered by event scheduling applications is that one has the possibility to change or even delete one's vote at any point of time. This is not necessarily possible in an easy way within a privacy-enhanced solution.

Changing a vote within the vote casting phase might help an attacker to undermine the privacy of participants (e. g., it may release release information about the secret key in homomorphic schemes). Completely deleting votes might make the selection process impossible as it may be the case that all votes are needed for the selection rule.

5 Conclusion

We formulated requirements of privacy-enhanced event scheduling and gave a short literature overview of existing solutions. Open research questions belonging to privacy-enhanced event scheduling where raised. Some of the research questions where already fulfilled in existing schemes (e. g., DCOP may solve complex selection rules and specify preferences; there is no motivation of legal-but-selfish votes when using e-voting schemes). However, to the best of our knowledge there is neither a scheme which is designed to be usable nor an application which offers usable privacy-enhanced event scheduling currently.

Acknowledgments. The author thanks Stefanie Pötzsch and Sandra Steinbrecher for their valuable suggestions.

The research leading to these results has received funding from the European Community's Seventh Framework Programme (FP7/2007–2013) under grant agreement № 216483.

References

1. Näf, M.: Doodle homepage (April 2010), http://www.doodle.com
2. Silaghi, M.C., Sam-Haroud, D., Faltings, B.: Asynchronous search with aggregations. In: Proceedings of the Seventeenth National Conference on Artificial Intelligence and Twelfth Conference on Innovative Applications of Artificial Intelligence, pp. 917–922. AAAI Press/The MIT Press (2000)
3. Yokoo, M., Hirayama, K.: Algorithms for distributed constraint satisfaction: A review. Autonomous Agents and Multi-Agent Systems 3(2), 185–207 (2000)
4. Léauté, T., Faltings, B.: Privacy-preserving multi-agent constraint satisfaction. In: Conference on Information Privacy, Security, Risk and Trust (PASSAT 2009) [34], pp. 17–25 (2009)
5. Modi, P.J., Shen, W.M., Tambe, M., Yokoo, M.: Adopt: Asynchronous distributed constraint optimization with quality guarantees. Artificial Intelligence 161, 149–180 (2004)

6. Maheswaran, R.T., Tambe, M., Bowring, E., Pearce, J.P., Varakantham, P.: Taking dcop to the real world: Efficient complete solutions for distributed multi-event scheduling. In: AAMAS, pp. 310–317. IEEE Computer Society, Los Alamitos (2004)

7. Mailler, R., Lesser, V.: Solving distributed constraint optimization problems using cooperative mediation. In: AAMAS 2004: Proceedings of the Third International Joint Conference on Autonomous Agents and Multiagent Systems, Washington, DC, USA, pp. 438–445. IEEE Computer Society, Los Alamitos (2004)

8. Franzin, M.S., Freuder, E.C., Rossi, F., Wallace, R.: Multi-agent meeting scheduling with preferences: Efficiency, privacy loss, and solution quality. AAAI Technical Report WS-02-13 (2002)

9. Greenstadt, R., Pearce, J.P., Bowring, E., Tambe, M.: Experimental analysis of privacy loss in DCOP algorithms. In: Proc. of ACM AAMAS, pp. 1424–1426. ACM Press, New York (2006)

10. Chaum, D.L.: Untraceable electronic mail, return addresses, and digital pseudonyms. ACM Commun. 24(2), 84–90 (1981)

11. Park, C., Itoh, K., Kurosawa, K.: Efficient anonymous channel and all/Nothing election scheme. In: Helleseth, T. (ed.) EUROCRYPT 1993. LNCS, vol. 765, pp. 248–259. Springer, Heidelberg (1994)

12. Ogata, W., Kurosawa, K., Sako, K., Takatani, K.: Fault tolerant anonymous channel. In: Han, Y., Okamoto, T., Qing, S. (eds.) ICICS 1997. LNCS, vol. 1334, pp. 440–444. Springer, Heidelberg (1997)

13. Abe, M.: Universally verifiable mix-net with verification work independent of the number of mix-servers. In: Nyberg, K. (ed.) EUROCRYPT 1998. LNCS, vol. 1403, pp. 437–447. Springer, Heidelberg (1998)

14. Jakobsson, M.: A practical mix. In: Nyberg, K. (ed.) EUROCRYPT 1998. LNCS, vol. 1403, pp. 448–461. Springer, Heidelberg (1998)

15. Cohen, J.D., Fischer, M.J.: A robust and verifiable cryptographically secure election scheme. In: SFCS 1985: Proceedings of the 26th Annual Symposium on Foundations of Computer Science (sfcs 1985), Washington, DC, USA, pp. 372–382. IEEE Computer Society, Los Alamitos (1985)

16. Benaloh, J.C., Yung, M.: Distributing the power of a government to enhance the privacy of voters. In: PODC 1986: Proceedings of the Fifth Annual ACM Symposium on Principles of Distributed Computing, pp. 52–62. ACM, New York (1986)

17. Sako, K., Kilian, J.: Secure voting using partially compatible homomorphisms. In: Desmedt, Y.G. (ed.) CRYPTO 1994. LNCS, vol. 839, pp. 411–424. Springer, Heidelberg (1994)

18. Baudron, O., Fouque, P.A., Pointcheval, D., Stern, J., Poupard, G.: Practical multi-candidate election system. In: PODC 2001: Proceedings of the Twentieth Annual ACM Symposium on Principles of Distributed Computing, pp. 274–283. ACM, New York (2001)

19. Chaum, D.: Elections with unconditionally-secret ballots and disruption equivalent to breaking RSA. In: Günther, C.G. (ed.) EUROCRYPT 1988. LNCS, vol. 330, pp. 177–182. Springer, Heidelberg (1988)

20. Fujioka, A., Okamoto, T., Ohta, K.: A practical secret voting scheme for large scale elections. In: Seberry, J., Zheng, Y. (eds.) AUSCRYPT 1992. LNCS, vol. 718, pp. 244–251. Springer, Heidelberg (1992)

21. Sako, K.: Electronic voting scheme allowing open objection to the tally. IEICE Transactions on Fundamentals of Electronics, Communications and Computer Sciences 77(1), 24–30 (1994)

22. Ohkubo, M., Miura, F., Abe, M., Fujioka, A., Okamoto, T.: An improvement on a practical secret voting scheme. In: Mambo, M., Zheng, Y. (eds.) ISW 1999. LNCS, vol. 1729, pp. 225–234. Springer, Heidelberg (1999)
23. DuRette, B.W.: Multiple administrators for electronic voting. Bachelor's thesis, Massachusetts Institute of Technology (May 1999)
24. Kellermann, B., Böhme, R.: Privacy-enhanced event scheduling. In: Conference on Information Privacy, Security, Risk and Trust, PASSAT 2009 [34], pp. 52–59 (2009)
25. Chaum, D.: Security without identification: Transaction systems to make big brother obsolete. ACM Commun. 28(10), 1030–1044 (1985)
26. Cranor, L., Cytron, R.: Sensus: A security-conscious electronic polling system for the internet (1997)
27. Herschberg, M.A.: Secure electronic voting over the world wide web. Master's thesis, Massachusetts Institute of Technology (May 1997)
28. Herlea, T., Claessens, J., Preneel, B., Neven, G., Piessens, F., Decker, B.D.: On securely scheduling a meeting. In: Dupuy, M., Paradinas, P. (eds.) SEC. IFIP Conference Proceedings, vol. 193, pp. 183–198. Kluwer, Dordrecht (2001)
29. Chaum, D.: The dining cryptographers problem: Unconditional sender and recipient untraceability. Journal of Cryptology 1(1), 65–75 (1988)
30. Diffie, W., Hellman, M.E.: New directions in cryptography. IEEE Transactions on Information Theory IT-22(6), 644–654 (1976)
31. Kellermann, B.: Datenschutzfreundliche Terminplanung. In: Mehldau, M.w. (ed.) Proceedings of the 26th Chaos Communication Congress, Marktstraße 18, 33602 Bielefeld, Chaos Computer Club, Art d'Ameublement, pp. 207–211 (December 2009)
32. Kellermann, B.: Dudle homepage (April 2010), http://dudle.inf.tu-dresden.de
33. Greenstadt, R., Smith, M.D.: Collaborative scheduling: Threats and promises. In: Workshop on Economics and Information Security (2006)
34. IEEE/IFIP: Proceedings of IEEE International Conference on Computational Science and Engineering, Conference on Information Privacy, Security, Risk and Trust (PASSAT 2009), Los Alamitos, CA, USA, IEEE/IFIP, vol. 3. IEEE Computer Society (2009)

Event Handoff Unobservability in WSN

Stefano Ortolani[1], Mauro Conti[1], Bruno Crispo[2], and Roberto Di Pietro[3]

[1] Vrije Universiteit, Amsterdam, The Netherlands
{ortolani,mconti}@few.vu.nl
[2] University of Trento, Trento, Italy
crispo@disi.unitn.it
[3] Università di Roma Tre, Rome, Italy
dipietro@mat.uniroma3.it

Abstract. The open nature of communications in Wireless Sensor Networks (WSNs) makes it easy for an adversary to trace all the communications within the network. If techniques such as encryption may be employed to protect data privacy (i.e. the content of a message), countermeasures to deceive context privacy (e.g. the source of a message) are much less straightforward. In recent years, the research community addressed the problem of context privacy. Some work aimed to hide the position of the collecting node. Other work investigated on hiding the position of an event—sensed by the WSN. However, the solutions proposed for events hiding either: (i) considered only static events; (ii) are not efficient. In this work, we describe open issues that we identified in the current research. In particular, we consider the problem of efficiently hiding mobile events.

1 Introduction

Due to the open nature of communications in Wireless Sensor Networks (WSNs), it is fairly easy for an adversary to trace all the communications within the network. If techniques such as encryption may be employed to protect data privacy (i.e. the content of a message), countermeasures to deceive context privacy (e.g. the source of a message) are much less straightforward. The research community has recently started addressing this issue in the unique context of WSNs [1]. The resource constrained environment, together with the enhanced adversary capabilities—e.g. the adversary can be mobile and eavesdrop all the communications—, have no correspondence in the wired setting, hence calling for novel solutions to address context privacy issues in WSNs [2].

In many applications (e.g., sensing and reporting the location of a convoy) the source of a message itself reveals the events a WSN is sensing. In order to protect these events, and thus assuring context privacy, we need to conceal that an event took place.

In this paper we identify the following open issues for the current research in this area:

J. Camenisch, V. Kisimov, and M. Dubovitskaya (Eds.): iNetSec 2010, LNCS 6555, pp. 20–28, 2011.
© IFIP International Federation for Information Processing 2011

- Open Issue 1. If bogus traffic is used to hide the real one, the adversary success (in capturing a node that routed a real event) might not be just linear with the amount of real event.
  - Open Issue 2. Solving the Open Issue 1 might imply losing the unobservability property.
- Open Issue 3. An enhanced privacy property, $\hat{k}$-anonymity, could be defined to describe the fact that an event cannot be distinguished between k-1 other events, but also requiring the other k-1 event to be not real events.
- Open Issue 4. Hiding the trajectory of an event should take into consideration the anatomy of the trajectory itself.
- Open Issue 5. The need for a common metric for privacy and energy consumption.

Organization. In Section 2 we revise the state of the art in the context of privacy preservation in WSN. In Section 3 we introduce the system model and the adversary model considered in this paper. In sections 4, 4.1, 5, 6, and 7 we present different open issues we identified in current research. We conclude our work in Section 8.

2   Related Work

Providing privacy to WSNs would allow a wider application of this technology. WSNs privacy has been first addressed from the point of view of communication confidentiality. Recently, also the context privacy has been investigated—a survey can be found in [2]. WSNs privacy also depends on the specific use of the network. E.g. [3,4] addressed the problem of *privacy preserving* data aggregation [5].

The problem of guaranteeing the privacy of a node sending data to the BS (not throughout the aggregation process) has been initially addressed leveraging how the messages are routed. In [6,7], the authors aim at hiding the source of a message, forwarding messages to the BS using random walks and dummy traffic. In [8,9] the BS location is protected by letting the nodes send messages to a random node, instead of the BS. This random node will then aggregate the traffic and communicate to the BS. This work consider a local adversary. In [10], the authors propose a partial k-anonymity solution for event source. While they also consider mobile events, the solution is quite energy demanding—property not desirable in WSNs.

In [11] the energy aspect has been taken more into consideration. The authors used *carefully chosen* dummy traffic to conceal the event source location and formalized the concept of unobservability for wireless communication. Nodes acting as aggregator proxies are used to reduce the communication overhead. Another solution involving dummy traffic but not proxies has been proposed in [12]. In [1] the authors demonstrate that to achieve perfect global privacy performance benefit must be sacrificed. They also introduce the notion of strong source anonymity. We observe that the solutions in [1,11] introduce a delay in the message reaching the BS. The solution proposed in [13] switches on demand from a statistically-strong source anonymity scheme (i.e. [1]) to a k-anonymity scheme (i.e. [10]). How to solve the handoff problem in a secure and distributed manner is left as future work.

Finally, randomizing the node ID has also been proposed [14]. However, the adversary model considered does not leverage techniques such as traffic and rate analysis.

3   System and Adversary Model

While previous solutions consider mainly static events, in this work we deal with mobile events. A mobile event involves a set of sensing nodes, in a way that is dependent upon time and location: the set of nodes sensing a mobile events define a handoff trajectory. Hiding the handoff trajectory of mobile events is not yet well investigated [13]. In particular, we deal with a specific type of mobile events: the ones originating on the WSN perimeter (i.e. its border) and eventually ending inside the WSN area.

The adversary is assumed to be global, passive, and external. Furthermore, we take into account the possibility that, once the traffic is analyzed, the adversary is willing to verify the gathered information. This means that the adversary physically checks the locations of both real and dummy events. Constrained by a given time interval T_a, the adversary inspects a subset of the nodes believed accountable for the traffic he previously gathered—the adversary's revenue is proportional to how many checked locations previously corresponded to a real event.

4   Open Issue 1: A Non-linear Adversary Gain

We want to guarantee the privacy of a real event being sensed by the WSN, considering the models described in Section 3. In particular, we aim to conceal which nodes sensed a real event. In order to this, we constrain other nodes to act as they sensed a real event too, thus sensing a dummy event. The property we want to guarantee can be summarized by the following definition.

Definition 1. *(Real event (T,k)-unobservability).* *We define a real event unob-servability over the variables T and k. In particular, consider the observation O for a time interval T. The probability of a real event e entering the network with a rate following a Poisson distribution of parameter l, $0 \leq l \leq k$, is equal to the probability of e given O. If this holds for each possible choice of l, then e is called (T,k)-unobservable. Formally:*

$$\text{if } \forall O, \; P(e) = P(e|O) \text{ then } e \text{ is } (T,k)\text{-unobservable.}$$

In other words, just observing the network does not give any information on the Poisson parameter l, $0 \leq l \leq k$ at which real events actually enter into the network.

Furthermore, we have the following aim. Let us assume that the adversary becomes active in tampering nodes—to check if these nodes sensed real events. We want to keep constant the adversary's success probability. In particular, we want to reach this target even if the rate of real events l varies.

A possible approach. In the literature real events are modeled as Poisson processes of a given ratio l [15]. Given this, a possible solution could leverage a self-adaptive scheme that, given real events taking place with rate l, carefully produces dummy events with rate $k-l$. The parameter k is the rate of the overall events (i.e. real and dummy ones) we assume the WSN can deal with. Hence, we select k such that the rate of real events l is $\leq k$.

If the overall rate k is fixed, the adversary is not able anymore to distinguish whenever a message corresponded to a real event. However, since the scheme would need to adapt to the ratio of real events, an additional parameter T is needed: this parameter defines the time interval in which every node "learns" the actual amount of real events. According to that, each node tunes the rate of dummy events. We call the so defined privacy property (T, k)-unobservability.

In this setting, with the growth of l, the success ratio of the adversary, that checks the event source location, grows as well since there are less dummy events. Given $\phi = k - l$ dummy events and l real events, a naïve solution would be to increase ϕ accordingly. In other words, we may want to keep a fixed ratio between ϕ and l. However, such a solution presents two different concerns: (1) the amount of events a WSN can generate is limited; (2) given a fixed displacement of the network, the more events there are, the more likely is that the adversary may discover a real event.

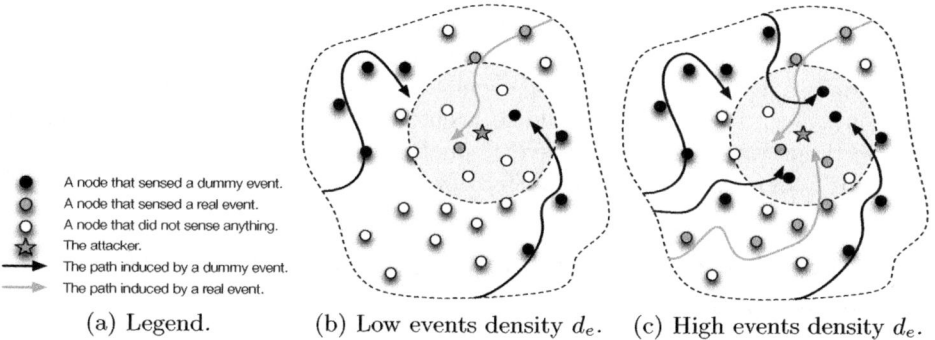

● A node that sensed a dummy event.
○ A node that sensed a real event.
○ A node that did not sense anything.
☆ The attacker.
→ The path induced by a dummy event.
→ The path induced by a real event.

(a) Legend.   (b) Low events density d_e.   (c) High events density d_e.

Fig. 1. The same WSN with two different events densities. The light circle represents the area the attacker may inspect in a given time interval T_a.

The former concern represents also the upper bound of real events the WSN is able to cope with. Since concealing real events requires us to generate additional dummy events, it is our interest to keep this quantity to a minimum. The latter concern is, in turn, depicted in Figure 1 (the legend is reported in Figure 1a).

Figure 1b shows a WSN where two dummy events (dark arrows) and one real event (light arrow) takes place, i.e. $\phi = 2$ and $l = 1$ (light arrow). The adversary, identified by the star, is able to inspect some of the nodes generating dummy traffic within the light circle. We remind that the adversary is constrained by a time interval T_a, therefore he may miss to find the depicted real event.

Let us analyze the following case now: we want to increase the dummy events ($\phi = 4$) in order to deal with a growth of real events ($l = 2$). Figure 1c shows this scenario: the area available to the adversary includes two different real events now. The probability that the adversary chooses the real event location intuitively increases: previously only one node out of eight would have been a successful choice (the node sending a real event). Merely increasing the amount of dummy events in a proportional manner is not the desired solution.

To address these concerns, we have to take into account the density d_e of the nodes relaying these events. Less events trivially correspond to a lower d_e. If we increase the events, the attacker has more nodes to check (i.e. higher d_e).

4.1   Open Issue 2: Proportional Adversary Gain Means Losing Unobservability

The solution reported in the previous section strongly relied on the concept of k-anonymity; since the amount of events was kept constant (k), l real events were concealed among $k - l$ dummy events. As long as k was not changing, real events hitting the network were made unobservable.

However, the more the real events were increasing, the higher was the probability of success of our adversary. To cope with that behavior we proposed to increase the rate of dummy events in a more than proportional manner. Unfortunately this countermeasure also enjoys a side-effect: a passive and global adversary has all the means to infer whether the overall amount of events k changes. This piece of information trivially discloses whether a WSN sensed more or less real events, hence real events do not enjoy the unobservability property anymore. Is it then possible to keep the adversary probability of success proportional while keeping real events unobservable? Is losing unobservability in favor of k-anonymity the right path to provide context privacy w.r.t. a WSN?

5   Open Issue 3: An Enhanced Privacy Property $\hat{k}$-anonymity

A subject is considered k-anonymous whenever it is concealed among $k - 1$ other subjects. This concept has often be applied to solve the problem of releasing sensitive data-sets [16]: a record was appointed as k-anonymous if the information for each person contained in the release (i.e. quasi-identifier) cannot be distinguished from at least $k - 1$ individuals whose information also appears in the release. Other work (e.g. [17]) applied the same idea to anonymous communication networks: a sender was considered k-anonymous in case an external observer was not able to distinguish which of the k peers was the actual sender.

However, since the adversary physically checks the locations of both dummy and real events, one might desire that not only one event cannot be identified within k possible events, but also that all the others k-1 events are bogus ones. We call this privacy property $\hat{k}$-anonymity. It holds if and only if a real event is concealed among $k-1$ dummy events.

6 Open Issue 4: Trajectory's Anatomy

In order to conceal to an adversary the location of nodes sensing an event, we outlined in Section 4 an approach that generates dummy events. These dummy events are generated by sending to the sink node the same type of message used whether a real event is sensed. A global adversary, in fact, can not distinguish between messages corresponding to real events and those referred to dummy events. The context privacy of a real event is therefore assured.

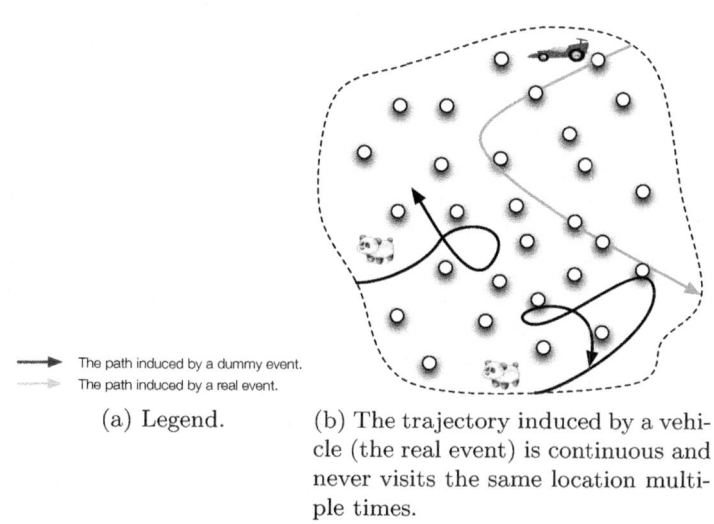

The path induced by a dummy event.
The path induced by a real event.

(a) Legend.

(b) The trajectory induced by a vehicle (the real event) is continuous and never visits the same location multiple times.

Fig. 2. Two different type of trajectories: a panda and a vehicle

However, since we deal with mobile events, we need to take into account which nodes are cooperating to build a dummy mobile event. In other words, since a mobile event is supposed to span across different nodes, a real event is expected to exhibit a trajectory. Therefore, whenever we generate a dummy event, a proper set of nodes must be chosen in order to produce a trajectory that resembles a real one.

This problem can be exemplified in Figure 2 by analyzing the trajectories produced by an endangered panda and by a vehicle. We do not expect a vehicle to visit multiple times the same locations; a panda instead is more prone to visit locations, such as source of water, which have already been visited. Trying to anonymize a vehicle with trajectories produced by a panda splits the set of sensed events in two different subsets: the subset of dummy events and the one of real events. Once the adversary obtains the opportunity to correctly label these two subsets, the privacy of real events can not be any longer assured. The type of events a WSN is supposed to deal with is therefore an important piece of information for any sound and secure solution.

7 Open Issue 5: A Metric for Privacy and Energy Consumption

Wireless Sensor Network is a novel field where the consumption of energy is often a system requirement. Moreover, since all the nodes have a physical location, the initiator and the recipient of any communication may become sensitive assets. A typical scenario is a WSN sensing a panda threatened by poachers: the node that initiates the communication trivially discloses the location of the panda; likewise, the recipient discloses where all the information is eventually collected. In both cases an adversary can be interested in the disclosed information.

A Privacy Enhancing Technology (PET) has an immediate impact on the battery life of any type of device. Data privacy, for instance, is often provided by means of some security primitives to encrypt and decrypt the exchanged data. No matter which primitives are chosen, the node's CPU becomes accountable of any additional computation. This means that any node sending or receiving encrypted messages will suffer from a reduced battery life.

However, as mentioned before, applications relying on WSNs advise for novel solution in the field of contextual privacy. Since what has to be concealed are the nodes taking part in a communication, any PET has to generate additional traffic to somehow anonymize the real communication. In case of an active adversary, we point out the following trade-off: the more dummy traffic is generated, the more private the real communication is; consequently the more energy must be consumed in the process.

Providing any sort of PET is therefore a rather expensive task. Existing works [18] proposed approaches to model the consumption of energy in terms of the adopted security primitive. What has not been yet proposed is a general model able to provide some reasoning to evaluate the proposed PETs. In particular, given a PET, we shall be able to assess the level of privacy the proposed solution provides, and the overhead produced in terms of energy consumption.

The latter is a rather interesting problem: since all the messages are eventually delivered to the sink node, we shall not expect an evenly distributed energy consumption. Instead, we may expect nodes close to the sink node to handle a higher rate of messages if compared to nodes lying on the perimeter. Figure 3 depicts exactly this behavior: nodes in the red circle are supposed to send more messages than nodes lying in a more external area. This kind of behavior creates what are known to be hot spots, i.e. areas where the consumption of energy is expected to be higher. We believe a PET shall take into the existence of these special ares, and mitigate their rate of occurrence.

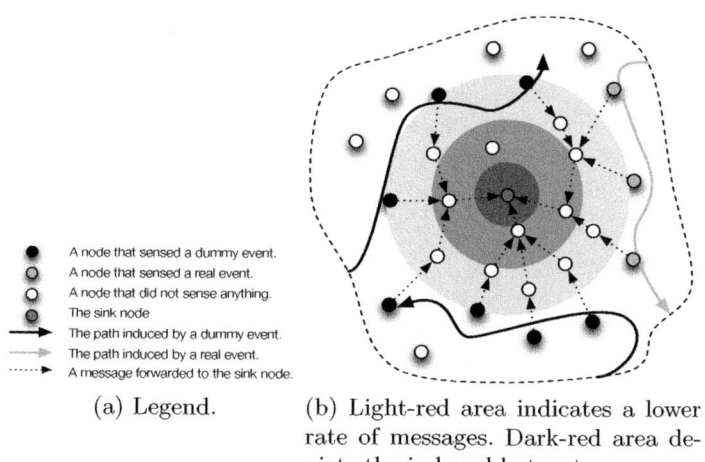

(a) Legend.   (b) Light-red area indicates a lower rate of messages. Dark-red area depicts the induced hotspot.

Fig. 3. Hot spot induced by the sink node being the ultimate recipient of any message

8   Conclusion

In this paper we described different open issues in context privacy of Wireless Sensor Networks (WSNs). In particular, we considered the problem of hiding mobile events—e.g. presence of animal or vehicles—that are sensed by the network itself. We observed how current solutions designed to hide static events either (i) are not able to hide mobile events, or (ii) are not efficient. We think that the solution of the open issues presented in this paper would remove barriers for a wide adoption of WSNs. Our future works aim to solve these issues.

References

1. Shao, M., Yang, Y., Zhu, S., Cao, G.: Towards statistically strong source anonymity for sensor networks. In: Proceedings of the 27th IEEE International Conference on Computer Communications (INFOCOM 2008), pp. 51–55 (2008)
2. Li, N., Zhang, N., Das, S., Thuraisingham, B.: Privacy preservation in wireless sensor networks: A state-of-the-art survey. Ad Hoc Networks 7, 1501–1514 (2009)
3. He, W., Liu, X., Nguyen, H., Nahrstedt, K., Abdelzaher, T.: Pda: Privacy-preserving data aggregation in wireless sensor networks. In: Proceedings of the 26th IEEE International Conference on Computer Communications (INFOCOM 2007), pp. 2045–2053 (2007)
4. Conti, M., Zhang, L., Roy, S., Di Pietro, R., Jajodia, S., Mancini, L.V.: Privacy-preserving robust data aggregation in wireless sensor networks. Security and Communication Networks 2, 195–213 (2009)
5. Xi, Y., Schwiebert, L., Shi, W.: Preserving source location privacy in monitoring-based wireless sensor networks. In: 20th International Parallel and Distributed Processing Symposium (IPDPS 2006), pp. 77–88 (2006)

6. Kamat, P., Zhang, Y., Trappe, W., Ozturk, C.: Enhancing source-location privacy in sensor network routing. In: Proceedings of 25th IEEE International Conference on Distributed Computing Systems (ICDCS 2005), pp. 599–608 (2005)
7. Ozturk, C., Zhang, Y., Trappe, W.: Source-location privacy in energy-constrained sensor network routing. In: Proceedings of the 2nd ACM workshop on Security of Ad hoc and Sensor Networks, pp. 88–93 (2004)
8. Conner, W., Abdelzaher, T., Nahrstedt, K.: Using data aggregation to prevent traffic analysis in wireless sensor networks. In: Gibbons, P.B., Abdelzaher, T., Aspnes, J., Rao, R. (eds.) DCOSS 2006. LNCS, vol. 4026, p. 202. Springer, Heidelberg (2006)
9. Jian, Y., Chen, S., Zhang, Z., Zhang, L.: Protecting receiver-location privacy in wireless sensor networks. In: Proceedings of the 26th IEEE International Conference on Computer Communications (INFOCOM 2007), pp. 1955–1963 (2007)
10. Mehta, K., Liu, D., Wright, M.: Location privacy in sensor networks against a global eavesdropper. In: Proceedings of the IEEE International Conference on Network Protocols (ICNP 2007), pp. 314–323 (2007)
11. Yang, Y., Shao, M., Zhu, S., Urgaonkar, B., Cao, G.: Towards event source unobservability with minimum network traffic in sensor networks. In: Proceedings of the 1st ACM Conference on Wireless Network Security (WiSec 2008), pp. 77–88 (2008)
12. Ouyang, Y., Le, Z., Liu, D., Ford, J., Makedon, F.: Source location privacy against laptop-class attacks in sensor networks. In: Proceedings of the 4th International Conference on Security and Privacy in Communication Networks (SecureComm 2008), pp. 1–10 (2008)
13. Yang, Y., Zhu, S., Cao, G., LaPorta, T.: An active global attack model for sensor source location privacy: Analysis and countermeasures. In: Chen, Y., Dimitriou, T.D., Zhou, J. (eds.) SecureComm 2009. Lecture Notes of the Institute for Computer Sciences, Social Informatics and Telecommunications Engineering, vol. 19, pp. 373–393. Springer, Heidelberg (2009)
14. Ouyang, Y., Le, Z., Xu, Y., Triandopoulos, N., Zhang, S., Ford, J., Makedon, F.: Providing anonymity in wireless sensor networks. In: IEEE International Conference on Pervasive Services, pp. 145–148 (2007)
15. Diaz, C., Seys, S., Claessens, J., Preneel, B.: Towards measuring anonymity. In: Dingledine, R. (ed.) PET 2003. LNCS, vol. 2760, pp. 184–188. Springer, Heidelberg (2003)
16. Sweeney, L.: k-anonymity: A model for protecting privacy. International Journal of Uncertainty Fuzziness and Knowledge Based Systems 10, 557–570 (2002)
17. von Ahn, L., Bortz, A., Hopper, N.J.: k-anonymous message transmission. In: Proceedings of the 10th ACM Conference on Computer and Communications Security (CCS 2003), pp. 122–130 (2003)
18. Potlapally, N., Ravi, S., Raghunathan, A., Jha, N.: A study of the energy consumption characteristics of cryptographic algorithms and security protocols. IEEE Transactions on Mobile Computing 5, 128–143 (2006)

Emerging and Future Cyber Threats to Critical Systems[*]

Edita Djambazova[1], Magnus Almgren[2], Kiril Dimitrov[1], and Erland Jonsson[2]

[1] Institute for Parallel Processing - BAS, Sofia, Bulgaria
{ead,kpd}@iccs.bas.bg
[2] Chalmers University, Göteborg, Sweden
{magnus.almgren,erland.jonsson}@chalmers.se

Abstract. This paper discusses the emerging and future cyber threats to critical systems identified during the EU/FP7 project ICT-FORWARD. Threats were identified after extensive discussions with both domain experts and IT security professionals from academia, industry, and government organizations. The ultimate goal of the work was to identify the areas in which cyber threats could occur and cause serious and undesirable consequences, based on the characteristics of critical systems. A model of a critical system is suggested and used to distill a list of cyber threats specific to such systems. The impact of the identified threats is illustrated by an example scenario in order to stress the risks and consequences that the materialization of such threats could entail. Finally, we discuss possible solutions and security measures that could be developed and implemented to mitigate the situation.

1 Introduction

Critical systems and networks constitute the critical infrastructure of society. The extensive use of Information and Communication Technologies (ICT) and their proliferation in many new areas, such as process control and critical infrastructures, pose substantial challenges to critical systems' security. Modern technologies are used for industrial process control and may introduce new vulnerabilities and even be the cause for incidents. On the other hand, advanced automation is widely used in critical infrastructures through industrial control systems, something that leads to new security problems. Critical infrastructures (CIs) themselves expand the scale of security threats with their complexity, large connectivity, interdependency, and possible cascading effects. The characteristics of critical systems thus highlight the need for security solutions *specific* to those systems and we feel that special attention has to be paid to those specific solutions.

In order to find possible solutions, we first have to identify and understand the emerging cyber threats to critical systems. One of the major objectives of the ICT-FORWARD Project[1] was to outline the critical threat areas where research

---

[*] This work was supported by the EU FP7 ICT-FORWARD Project under Grant agreement no. 216331/14.09.2007.
[1] http://www.ict-forward.eu

J. Camenisch, V. Kisimov, and M. Dubovitskaya (Eds.): iNetSec 2010, LNCS 6555, pp. 29–46, 2011.
© IFIP International Federation for Information Processing 2011

efforts have to be invested and countermeasures have to be devised. After a series of discussions with domain experts from industry, academia, and government organizations, a survey of emerging and future security threats was prepared [1].

This paper summarizes some of the identified security threats to critical systems and discusses the open research problems in applying security mechanisms to such systems and in developing new solutions.

2 Critical System: Modelling and Specifics

In our work, we used the following definition of a threat:

Definition of a threat: A threat is any indication, circumstance, or event with the potential to cause harm to an ICT infrastructure and the assets that depend on this infrastructure.

Our version is related to a variety of other definitions that exist in the literature, such as the ones provided by *ISO/IEC* and the *EU Green Paper for Critical Infrastructure Protection* [2]. In both these cases, a threat is described as an event, circumstance or incident that has the potential to cause destruction or, more generally, harm to the system or organization that is exposed to the threat. We adapt our definition to explicitly refer to ICT infrastructures and assets, as this is the scope of the FORWARD project. However, we observe that the definition is reasonably general to accommodate a wide range of possible threats and scenarios.

In order to be able to focus our efforts, we further developed a model of a critical system to help distinguish the most interesting and pressing security threats.

2.1 Modelling a Critical System

In Fig. 1, a generic ICT system is shown. It is defined as any system that delivers service to a group of users. Such a system is subjected to a number of threats, which may influence the service delivery to the users. We could leave the *system* box as a black box in the diagram. A ranking of the most important emerging threats can still be performed on such a black-box system. However, by knowing more about the system box, we can better judge what types of emerging threats will be the most severe and therefore important.

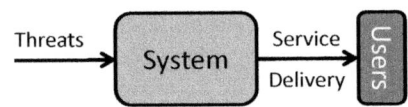

Fig. 1. A model of a generic ICT system

The paper focuses on the ICT that supports critical infrastructures. Understanding the emerging threats to such systems is important because the consequences can be very dire. In our study, we consider the system from Fig. 1 to be a critical system.

Definition of a critical system: *We define a critical system (CS) to be a system that delivers a critical service to a group of users. A critical system consists of a traditional critical infrastructure or a critical application and supporting Information and Communication Technology.*

More specific definitions of a critical system can be found elsewhere in the literature. Here, we by purpose refrain from a very specific definition. The criterion of criticality may change over time and each professional group that discusses the issue has their own definition. By presenting a general model with the important salient properties found across many critical infrastructures, we can focus on the issues that will remain relevant in the future in our discussions.

In Fig. 2, we have expanded the *system* box shown in Fig. 1. Assuming a critical service delivery, we can further detail the structure of the *system* box based on the model of such critical systems.

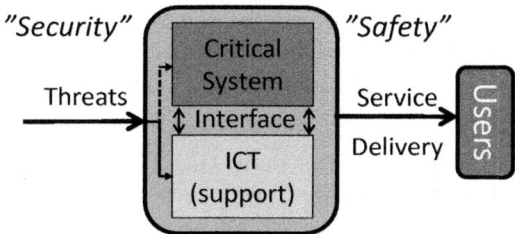

Fig. 2. A model of a specific system for critical services

We would like to emphasize that there are four boundaries in Fig. 2: the outer system boundary, the two inner boundaries to the critical system (the dashed line) and to the supporting ICT system (the flat line), and the boundary (interface) between the critical system and the ICT system. The threats can then be divided into four groups according to the boundaries shown in Fig. 2.

1. Threats targeting the whole CS–ICT.
2. Threats targeting the interface between the CS and ICT.
3. Threats targeting the ICT part.
4. Threats directly targeting the CS.

We do not consider threats that directly target the critical system (4), in that it is a critical infrastructure. Such threats are already discussed and accounted for in other working groups in the EU and elsewhere. The focus here is on cyber threats, often directly targeting the ICT structure by their very nature. Thus, the focus is on problems related to the supporting ICT infrastructure, that is (1) – (3) in the list above. We would like to emphasize (2) in the list, as the particulars of this interface may be prone to many security vulnerabilities.

2.2   Specifics of Critical Systems

In order to identify the specific security problems in critical systems we developed the described *model of a system for critical services*. Based on that model we can outline the security specifics of critical systems and why the threats to critical systems and the respective countermeasures need a different approach compared to that of more traditional systems. Even though the emerging and future cyber threats seem common for all ICT applications, there are specific issues regarding the sub-domain of critical systems. There are many differences compared with a regular ICT system. In a regular system, there is no ICT–CS boundary. A regular ICT system is not normally connected to a system governed by physical laws. This implies that a regular ICT system does not have the same constraints in terms of timely input of data or a similar limitation on the types of interfaces available. No critical service is delivered by a regular system.

We divided the *system* box in Fig. 2 into two parts: one part being the actual critical system (or critical infrastructure) and the second part being the supporting ICT infrastructure. In some cases, the critical system is of an ICT nature; in other cases it is a traditional process control system. Some of the specific characteristics of a critical system as shown in Fig. 2 are described in more detail below.

Critical service. A critical system is delivering a critical service to users, which has to be preserved and maintained even in the case of cyber attacks. The disruption of operation of such systems will lead to severe consequences.

Complexity and availability. The complex architecture of critical infrastructures hampers investigation and assessment of the impact of threats. Further complicating the issue is that many of these systems need to run around the clock all days of the year, meaning that a system cannot simply be brought off line for testing or security update.

Many and different interfaces. There are various types of interfaces to a critical system, since it is the result of combining several independent systems and they differ greatly in many ways. This affects the vulnerability of the system as a whole. Critical systems have specific and diverse relations with ICT systems and between internal systems. Further, the system mixes interactions of human operators (slow response) with computer services (fast response) through a variety of interfaces. Many times these interactions are rather complicated in that the access modes vary and the time frames between the parts are widely different.

Interdependency issues (long chains of dependencies). One of the important issues for critical infrastructures is the interdependencies among the infrastructures. There may be long and complex dependency chains. An attack against any of the services may cascade unpredictably through the system. In [3], the role of ICT in critical infrastructures is defined with the term cyber interdependency. An infrastructure has cyber interdependency if its state depends on information transmitted through the information infrastructure.

Data is important. Almost always, data is important [4]. This is especially true for *financial services*. It is also true for other types of systems, such as air traffic control, where data are underlying even the simplest decisions.

An underlying physical process. Many times, a physical process is underlying the critical system. The system has to observe time constraints which are hard to combine with certain security measures. The critical system may be part of a *control loop* in the physical system. Thus, critical systems have physical and possibly a very complex interaction with the environment. Security functions integrated into the critical system must not be allowed to compromise the normal functionality of the critical system [5].

Real-time constraints. The connection of a CS to the physical world implies that critical systems are often real-time, as they are determined by physical systems. They may also be considered real-time in that they deliver a critical service that should not be interrupted. Depending on the specific system, the term "real time" may imply very different time scales – from seconds to days. Critical systems are generally time-critical and have to respect some acceptable levels of delay and jitter dictated by the individual installation. Some systems require deterministic responses. This may mean that they have to observe time constraints, which are hard to combine with certain security measures. High throughput is typically not essential to CS. In contrast, ICT systems normally require high throughput, and they can typically withstand some level of delay and jitter [5].

Many owners, policies, and domains. Often, a critical system has many owners and this fact is emphasized through the deregulatory nature of policy decisions taken lately. The mixed ownership affects the system as a whole, in that there are artificial interfaces between the parts and each part may be governed by its own security/safety policy. For example, data is often sent over both propriety networks and the Internet.

Trade-off between safety and security. Based on the tradition of safety-critical systems, safety is and has been emphasized over security. For example, passwords are sometimes avoided by intent; it is reasoned that sometimes it is very important to immediately be able to control a process (to stop it from reaching a critical point), and a password would only slow down the operators. Thus, no regards to integrity or access control exists in such a system and such features cannot easily be added later, or added to one part of the system if another part lacks such support.

Mismatch of practices between CS and ICT systems. Operating systems and applications in critical systems may not tolerate typical IT security practices. Legacy systems are especially vulnerable to resource unavailability and timing disruptions. Control networks are often more complex and require a different level of expertise (e.g., control networks are typically managed by control engineers, not IT personnel). Software and hardware are more difficult to upgrade in an operational control system network. Many systems may not have desired features including encryption capabilities, error logging, and password protection [5].

The human factor plays a pivotal role for proper operation. The human being is considered to be the weakest point in a critical system. The roles include operators in control rooms, engineers taking technical decisions, managers and decision-makers for future strategy development. On the other hand, insiders with experience of and knowledge about the critical system could be a serious threat as seen, for example, in [13].

3   Example Scenario

We illustrate with an example scenario how the specific characteristics of critical systems described above influence information security of these systems. Although the example scenario is completely fictional, it was considered by domain experts as being realistic in that it shows some real or emerging threats for which no ready solutions are available.

The example scenario takes place on an oil platform. That domain, i.e. the oil platform, is specifically chosen because it embodies many of the problems related to critical systems' security highlighted by domain experts, such as the trend to increase efficiency through greater automation and more remote operations. The organizational structures are becoming more complex and there is an increased reliance on computer systems that are vulnerable to malfunctions and malicious attacks. The introduction of ICT opens up the previously isolated critical infrastructures to the information infrastructure and exposes them to threats.

The antagonist in the scenario is a single person with strong environmental ties who wants to make a "statement" about the dangers of the current oil dependency by shutting down an offshore drilling platform. Using his skills and some internal knowledge about the system, he manages to reach the main functionality of the platform. However, instead of stopping the production of oil on the platform, his attack causes a large oil spill with a severe environmental impact. Below, we give a brief overview of the attack, which is further elaborated in [1].

3.1   The Attack and Its Consequences

In order to reduce costs and increase effectiveness, oil platforms are connected in a bigger infrastructure involving remote control centers and a number of expert nodes in case problems occur. The antagonist in the example scenario takes advantage of such a trusted expert node by installing malicious software that does not spread actively, but propagates only when the victim host initiates its own connection to this server, thus working in much the same way as current malware in the web domain. The malicious server software is tailored towards the victim environment. The antagonist has an insider's view (from his previous work on the oil platform) and knows its weak points, i.e., vulnerable routers that can be corrupted, as well as the means to grant himself sufficient authorization for his goal of shutting down the oil production of the platform (known passwords). He manages to reach the control network and connects to a critical console. At that point, the safety systems on the platform fail and thousands of tons of oil flow into the ocean.

3.2   Related Threats

In Sect. 4, we go through the threats against critical systems in detail. Here, we highlight the threats particularly important to the execution of the attack described in the scenario. Issues related to the *use of commercial-off-the-shelf components*[2] and *retrofitting security to legacy systems* are fundamental to this scenario. Some of the systems are simply vulnerable to "normal" malicious code, and the antagonist uses this fact to download his code onto the offshore platform. The antagonist then uses his detailed knowledge of the system (*the insider threat*) for his next step in the attack. As *safety takes priority over security* in many industrial domains, we emphasize the non-existing password policies at the offshore platform. The domain is complex and issues such as *unforeseen cascading effects* and other problems due to scale also work in favor of the antagonist. Even though not explicitly mentioned, the human factor probably played a role in this scenario, and better designed *user interfaces* might have alerted the operators in time to the failure of the safety system. Finally, we would like to point out that in this story the antagonist was never brought to justice; legacy systems, where security has not permeated the design, seldom have the necessary sophistication for allowing advanced forensic analysis.

3.3   Discussion

Some of the discussed related threats are common to all ICT systems and there are countermeasures developed to overcome them. The problem with critical systems is that the security techniques cannot always be implemented directly. In most cases, they need to be modified according to the timing and performance requirements of the critical system. Even if tailored to the system's constraints, security mechanisms may still not work properly, since it is almost impossible to test them before implementation, because critical systems' operation should not be interrupted.

More interesting for our study, however, are those security threats that are specific to critical systems. Some of them are known to the professional community and there are solutions proposed and implemented. Other threats are just beginning to appear with the introduction of modern technologies in critical systems and the new, not studied or poorly understood, interactions of the CS and the ICT. We will focus our discussion in the next section on these specific security threats.

---

[2] The threat names are printed in italics.

4   Emerging and Future Threats to Critical Systems

Based on the characteristics of critical systems discussed in Sect. 2.2, we have identified areas where security threats might grow in the future and where new solutions should be sought for. The identified threats to critical systems summarize the views of many experts both from the domain of information security and industrial automation/critical infrastructure protection, and reflect the general vision that critical systems can become an attractive target for cyber attacks and that the cross area of ICT and CS is an open field for security research. Fortunately, there are not yet many (publicly known and documented) examples of successful attacks to critical systems, but the present experience shows the clear need for effective and specific countermeasures in this domain.

In the following subsections, we describe the identified cyber threats to critical systems and discuss some of the possible solutions. We focus on the ones relevant to the described scenario, but also include other important threats to critical systems. The full list of identified threats can be found in [1].

4.1   Use of Commercial-Off-The-Shelf Components

Threat. The use of Commercial-Off-The-Shelf (COTS) components and systems can make any system, but especially a system connected to a critical infrastructure, vulnerable to a variety of attacks. There are two problems with COTS components. The first problem is related to hidden functionality and outsourcing, as described in Sect. 4.9. The designer has no real control over the product he is introducing into his system. The COTS product is designed (and manufactured) elsewhere and the documentation can be incomplete or even faulty. There is no guarantee that there is no hidden functionality. Nor can its absence be verified, as discussed in Sect. 4.9. The second problem is related to the generality of the COTS systems versus the sometimes very specific requirements of the environments where they are used. It is this second problem we describe below.

To reduce cost and time for design, the use of COTS systems and components in critical applications seems attractive and will thus continue. COTS systems are often used in industrial automation process-control systems because they are cheaper to deploy and may include more functionality than a custom-built system. However, there is a gap between the priorities (safety versus cheap COTS components) and this gap leads to new challenges to security and reliability. For example, COTS systems are prone to "normal" virus infections and attacks, so attackers do not need to specifically tailor their malicious code to these systems. There will be remote access through connections to the Internet, leading to new threats. Response management is needed, coping with incidents – recovery, isolation, and restoring the system to a working state. Forensics should also be applied to determine the responsibilities.

There are some projects (e.g., DEAR-COTS [22]) where COTS components are applied to design distributed computer-controlled systems. They are organized using redundancy and design diversity to make the system dependable and secure. Some of the issues addressed in DEAR-COTS are the use of emerging information technologies to cope with heterogeneity issues while providing a dependable user-friendly man-machine interface.

Possible solutions. No good solution exists, but various work-arounds, such as using COTS systems with some fault-tolerant approaches (replication, diversity approach); applying COTS components in non-critical areas only; introduce and manage heterogeneity; or use of a compact and trusted application base.

Another possible approach is to introduce semantic technologies, i.e., to take a holistic approach to security with semantic technology (e.g., service-oriented architecture). Physical components should be classified, as they have to be defined from the basis. We have to identify and decide what and how to protect, i.e., an assessment of the assets to be protected has to be done.

4.2   Retrofitting Security to Legacy Systems

Threat. Security can seldom be retrofitted to an existing system, but due to economical constraints this is sometimes considered necessary. Most critical systems are created to provide a certain functionality. Safety and control characteristics are the natural focus of such systems. Thus, applying security measures afterward instead of incorporating them in the original design could constitute a problem. For example, the in-vehicle network has historically been a closed environment responsible for the control and maneuverability and safety of vehicles. The in-vehicle network has been designed to provide this functionality and security has not been part of the design. In the connected car of the future, external communication is allowed to interact with the previously isolated in-vehicle network. Thus, the in-vehicle network is opened up to potential attacks. Designing security solutions for the existing in-vehicle network creates difficulties as real-time constraints, protocol and hardware limitations need to be considered. In addition, security solutions must not interfere with the functionality provided, e.g., by imposing delays as this could have serious consequences from a safety perspective. Due to economical constraints it may not be possible to redesign the entire system with security in mind. Either the best possible security solutions considering the existing system are developed and applied and as a result possibly degrading the system's performance, or good enough solutions are applied to ensure that the existing system's functionality is left unaffected.

Possible solutions. The short-term solution could be a better understanding of how to best adapt security to such systems. Experts recommend [15] to study all connection points in the network, understand what traffic has to flow from the old networks into the business network. The information should be flowed through a more modern server, which can be better protected and analyze the traffic in real time. In general, analyzing the current architecture in detail and cataloging all software running on the control networks help discovering the weaknesses of the network and strengthening its security.

New architectures can be developed where security permeates all parts of the design for the long term. Migrating to new technologies, however, takes time, while security is needed at the present moment and this reality could influence the process of introducing new and more secure technologies.

4.3   The Insider Threat

Threat. A definition of the "insider threat" is given in [13]. This threat lies in the risk that a trusted employee betrays his employer by conducting some kind of malicious activity. Insider betrayals comprise a broad range of actions, from theft or subtle forms of sabotage to more aggressive and overt forms of vengeance, sabotage, and even work place violence. Insider activities cause financial losses to organizations, have negative impacts on their business operations, and damage their reputation.

In [13], it is argued that the nature and seriousness of the threat requires a combined view of physical and IT security systems and policies. Although physical and cyber threats from insiders manifest differently, the concepts are quickly converging as many potential attacks bear characteristics of both physical and IT sabotage, fraud, or theft.

Some interesting results from a study on the insider threat [14] show that a negative work-related event is most likely the trigger to most insiders' attacks. Furthermore, the majority of insiders planned their activities in advance. An observation is that the majority of insiders were granted privileged access when they started work, although less than half of the insiders had authorized access at the time of the incident. An interesting point is that both unsophisticated and relatively sophisticated methods for exploiting a system's vulnerabilities were used. Remote access was used to carry out the majority of the attacks. Many times, the insider attacks were only detected when there was a noticeable irregularity in the information system or when a system became unavailable.

Possible solutions. Effective strategies for discovering an "insider" is an open research question. The recommendations from [13] include low-cost, easily implemented policy solutions for near-term effect: education and awareness, employee screening, technology policy, information sharing. In the long-term aspect, further guidance, findings, samples, and tools are needed. Some solutions for IT systems/cyber security could be the following: to use integrated IT and physical security system tools to identify rule violation patterns for potential insider threat behavior; to use dual protection access technologies (e.g., biometric, key card or encryption key verification); to use dual control access mechanisms to protect high-value systems and processes; to manage access, integrity and availability of computer systems (e.g., identity management system). Control over creation and termination of user and administrator accounts and maintaining security/access rights should be done by segregation of duties. Using data loss prevention tools could help stopping the leakage of information outside the network and can be a measure to detect an insider activity.

4.4   Safety Takes Priority over Security

Threat. In the domain of critical systems, both safety and security are important but in certain scenarios, safety takes priority. Based on the tradition of safety-critical systems, safety is and has been emphasized over security. Giving priority to safety, however, is not just a traditional vision. It is justified by the potential losses after a safety incident. Safety of critical systems is important because of critical system's interaction with the physical world and the possible risks of that interaction. Security is usually considered being of less importance compared to the major safety issues of the actual CS. With the extensive use of ICT in critical systems, however, security should be considered more seriously, since security and safety are very interrelated. Problems with security can lead to safety issues. Thus, a security attack can lead to a safety problem and endanger lives.

Complicating the issue is the fact that control system professionals are often not aware of security risks, since these are not considered part of the normal system operation. The emphasis in control systems is on safety and availability aspects. On the other hand, IT security specialists use known techniques from a normal ICT system to introduce security, but may be missing important safety and control characteristics of the specific CS, as discussed in Sect. 2.2 and Sect. 4.2. This lack of mutual understanding between the control and security communities makes the overlooking of security a problem. Control specialists and even the management personnel of organizations are security-unaware and tend to neglect security measures and tools. Sometimes people with little experience or with different primary tasks operate the supporting IT system and they are more prone to do mistakes or ignore security alerts.

Possible solutions. As we stated previously, the understanding that safety and security are interrelated is of very high importance and will lead to improvement in overall security and safety policy. A better understanding of the domain for the IT security experts is necessary. On the other hand, the control community should be aware of the important role of security measures to safety. Work should be done on changing the mindset. Some simple technical measures could be to document changes done to the system in order to facilitate the implementation of security tools where they are most needed; keep the control traffic off the business network; document all the software installed on the network, etc.

4.5   Unforeseen Cascading Effects

Threat. Interconnected systems and networks are difficult to model properly and interdependencies between them can lead to cascading effects that are hard to foresee. This is due to the inherent complexity of the connected systems. It is claimed that nobody *really* understands a network such as the Internet anymore, nor even many smaller interconnected, heterogeneous networks that have been deployed over the past decades. Further, testing is virtually impossible due to the complexity and scale. In particular, testing is often impossible when the system is connected to a critical infrastructure with real-time requirements.

An important class of cascading effects occurs when, e.g., some section of the Internet is attacked or overloaded to the point of service denial and another (perhaps critical) system depends on that section. Even though the attack was not directed against the critical system per se, it is affected indirectly.

It is clear that dependencies are responsible for unforeseen cascading effects. Unfortunately, dependencies in large networks and systems are very difficult to understand due to their complexity. Even though system complexity is an issue in many areas, some factors related to critical systems make the issue of the complexity extra severe in such environments. First, due to the deregulation of markets, critical infrastructures are often run by different organizations that need to cooperate. These organizations are seldom a single unit, but they are comprised by many smaller units as virtual organizations. A complicating issue is then that part of the system may be governed by proprietary protocols while others use open standards. Different system owners may not trust each other, and different parts of the system may be governed by their own safety/security policies.

Possible solutions. What we need are new, more appropriate modelling tools and an overall better, probably structured and hierarchical, architecture with a security baseline. Removing the human from the loop and introducing automation may help. On the other hand, the seemingly intuitive action scripted in automated systems might be completely wrong in certain systems and lead to large problems. For essential services, it is important that dependencies should be tracked from the design phase onwards.

4.6   User Interface

Threat. The human plays various roles in control systems at all levels of their operation. For example, in a real critical system, it has been estimated that in some situations, human reliability can fall from 10^{-4} to 10^{-3}, whereas a system's reliability is maintained at 10^{-9}. Incorrect interactions with the system, handling other operator errors, and complex interdependencies as described above make it difficult to correctly work with the system. For these reasons, the human being is a serious factor when considering overall system security.

It is imperative to wrap new solutions to upcoming and even existing threats in understandable and discreet user interfaces to make sure they are properly used. The user information overload is a constant problem that is very likely to persist for a long time and hinder solutions for security problems to catch, even if they already exist.

Possible solutions. The education and training of personnel working in critical systems is a constant task that can help maintain an up-to-date knowledge on systems and networks. The awareness of security risks should be raised. There are many bad practices (e.g., running un-patched versions of software, using default configurations and passwords, etc.) that could easily be removed by making people understand the role of security measures. A sound and evolving security policy in the organization is needed to mitigate security risks. There are approaches to model the user (cognitive modelling) and user-interactive properties that could be used to improve the interaction of the users with the systems.

Another approach is to model and design the systems in such a way that they are more easily comprehended and understood. This would include, e.g., structural design, encapsulation, intuitive interaction interfaces, etc.

4.7   Sensor Networks

Threat. The convergence of control with communication and computation will make sensor networks the new dominant "computing class." This class will provide the ability for large numbers of interconnected sensors, actuators, and computational units to interact with the physical environment. This computational shift is going to bring a big shift also on computer security issues.

One problem is that small sensors require a means to communicate. This is typically a wireless connection. However, in addition to the security concerns of wireless networks in general (discussed in Sect. 4.8), wireless sensor networks have a number of additional issues. For example, the nodes in sensor networks are in general very limited in terms of battery, storage, and computational power. Therefore, strong cryptography and other general security tools are of limited use, if at all available. An attacker can have much more powerful hardware than the nodes being attacked. Sensor networks also typically reside in unattended environments where an attacker can physically destroy nodes, add malicious nodes or in other ways tamper with the hardware of the network. It is usually hard to distinguish the natural failures of the nodes in a sensor network from a malicious attack where nodes are deliberately destroyed.

There are many venues of attacking sensor networks [6,7], including the following: snooping information, inserting false or misleading information, jamming radio channels, making nodes run out of battery by never letting them sleep, giving the impression of phantom nodes that do not exist, giving the impression of connectivity that does not exist, making messages go through an attacking node that can selectively drop messages from the system.

Possible solutions. In summary, we consider the following three approaches worth further pursuit in the context of sensor devices.

- Autonomic solutions where the system will continuously evolve and control its security.
- Solutions that will mask subsystem takeover.
- Combining sensor information with physical information for verifying certain operations.

4.8   Wireless Communications in Industrial Environments

Threat. Wireless communications offer many convenient advantages compared to traditional wired communications within the industrial domain, such as:

operator mobility, safety by enabling remote access to noxious environments, access security for visualization and optimization, and the immediate benefits of their deployment [9].

Today, wireless communications are not yet widely used in practice in industrial environments. Most plants are only considering them for information gathering in the form of measurements, but not for closed-loop control [8]. Based on their advantages, however, a greater adoption of wireless communications in industrial control can be expected, thus with an overall growth in their market share. Experts from WINA and ISA [10] predict that within 10 years, even critical control communications will be wireless.

Recently, following the WirelessHART and ZigBee Alliance announcements and after approving the SP100 standard for industrial wireless communications by ISA, there is already use of wireless communications in industrial and even critical applications. Despite this, the single industrial wireless standard ISA-SP100.11 does not give enough guarantees for dependability and security to critical systems and applications.

One main security aspect of the wireless communications in general follows from the unbounded nature of radio frequency propagation. The perimeter of a wireless network cannot be limited and controlled as can be done with a wired network. There are reflected signals, which find their way out of buildings. These dispersed signals could be detected by motivated attackers that could then attempt to interfere with them if they are in physical proximity of the facility. Thus, traffic can be passively captured and an attempt to penetrate the network could be made with the aim to reach other connected enterprise networks.

Possible solutions. The first and main consideration when addressing security of industrial wireless communications is the conformity to the ISA-SP100 Usage Classes. Many useful and detailed recommendations for securing wireless networks are given in [11].

The IEEE 802.15.4 standard [12] gives some recommendations how to use guaranteed transmission mode and secure mode. It is shown that cryptographic randomization, agility, and diversification, in a game-theoretic context can provide the tools for building resilient wireless networks against both external and internal attacks. Such techniques can even allow the identification of internal attackers.

4.9 Hidden Functionality

Threat. One threat of paramount importance is that of hidden functionality in systems, and in particular, in software. Hidden functionality may comprise almost any functionality, but common examples are back doors, i.e., secret and undocumented entries to a system, and Trojan horses. Such functionality can be introduced into the system by accident, but the most common reason is that somebody, for example, the designer or maintenance engineer, enters this functionality for his own, in many cases malicious, purposes. In other cases, it is introduced for commercial reasons. Regardless of its purpose, the idea is that this extra hidden functionality is not known by the authorized user and the rightful owner of the system.

It is evident that such functionality presents an enormous threat. Not only is it unknown, but it is also put into the system in such a way that it is very hard to discover. Furthermore, this functionality is uncontrolled and can lead to a large range of very detrimental impacts on the system. As an example, in the U.S., the possibility of malicious hardware used for espionage, or even for terrorist activities is considered an emerging threat. Most hardware fabrication is nowadays outsourced. Circuits can be added on chips at the fabrication plant to offer a back door to potential attackers, or perform some other action. It is technically very hard for vendors to detect whether the produced hardware follows their design to the letter.

Possible solutions. It is very difficult to find solutions to this problem. Any type of remedy would imply the ability to prove, or at least make plausible, that no such functionality exists. Unfortunately, there are significant theoretical obstacles in proving the *absence* of something. It is certainly possible to find and remove such functionality, but to verify that there is none left after removal is extremely hard. Still, the only possible solution would be to develop better validation and verification methods and tools. A methodology for measuring security could be one of them as well as runtime detection of any unknown (malicious) functionality. In the short term, potential solutions to this problem might involve the use of secure and trusted fabs for critical hardware, such as the one used for aviation and military equipment.

4.10   Next Generation Networks

Threat. Recently, there is a general trend for carrying multimedia in the field of electronic communications. Under the pressure of the Internet, on the one hand, and because of the increased service requirements of end users, on the other, some telecommunication companies are migrating to the so-called Next Generation Networking (NGN).

NGN is a broad term describing some key architectural modifications in the telecommunication core and access networks that have been deployed in the last five years. The main goal of NGN is that one network transports all information and services (voice, data, and multimedia) by encapsulating them into packets. NGNs are commonly built around the Internet Protocol and therefore the term *all-IP* is also sometimes used to describe the transformation towards NGN [16].

The openness and easy access and usage of NGN lead to an increased number of vulnerabilities and extreme attention to security measures must be paid. Recently, many security experts bring up the attention to the specific vulnerabilities of the NGNs. The most exploited among them are [17,18]:

- Knowledgeable end users can gain access to the control plane of "all-IP" networks like NGNs.
- Large number of external connectivity points (and from any other point/site of the Internet)

- Shared core network among several NGN operators (the possibility of occurrence and the variety of vulnerabilities is higher)
- Malicious users can manipulate the traffic more easily as no physical access is required.

More than 32 fundamental vulnerabilities in NGNs are described as a result of the systematic assessment of NGN vulnerabilities [19].

Possible solutions. Security mechanisms on open packet networks will be very different from those of legacy telecommunication services in many aspects. In legacy networks, being circuit-oriented vertical networks, much policy management was "built into" the integrated service, comprising all aspects of the network. Security will need to be addressed differently in the NGN. The design and implementation of NGN need to meet complex requirements, which complicates its security architecture. As a consequence, it is difficult to use a single standard to define it [20]. As a present security solution it was recommended in [21] to use *multiprotocol label switching* (MPLS) virtual private networks to construct an NGN virtual private bearer network, and thus logically separate NGN services from traditional data services. As telecommunications companies already deploy NGNs in different forms (e.g., Vivacom in Bulgaria, KPN in the Netherlands, Ireland [16], British Telecom's *21CN*), this is an important problem.

5   Conclusion

Although information security, as a fast developing research direction, offers new solutions to counter cyber threats, there are domains where the existing security techniques cannot be applied directly. In some areas, the necessity to protect systems from cyber attacks is just beginning to be realized. Even though there is a rising interest and concern of the lack of cyber security of critical systems, the research in this area is still scattered and somewhat isolated to particular domains. A more thorough understanding of the risks and the need for new security solutions that focus on the emerging threat areas and the specific characteristics of critical systems is necessary.

The paper demonstrates some of the problems in implementing security in critical systems. The identified and described threats to these systems indicate the research areas where new security solutions are needed. Current practice shows that known IT security measures should be implemented considering the specifics of critical systems. Those measures should respect time constraints, continuous operation mode of these systems, their requirements for availability and safety, their heterogeneity, complexity, and interdependence, etc. New, possibly holistic, solutions should be developed, e.g. building in security at design level, applying service-oriented architecture, organizing systems according to the "defense-in-depth" strategy, resilience approach to their design and operation.

We have to note that critical systems encounter many different security problems that are a mixture of technological, psychological, and social issues. This calls for interdisciplinary approaches to be adopted to address the diverse threats to critical systems.

References

1. ICT FORWARD Project: Deliverable D3.1: White book: Emerging ICT threats (2010),
   http://www.ict-forward.eu/media/publications/forward-whitebook.pdf
2. Commission of the European Communities: Green Paper On a European Programme for Critical Infrastructure Protection (2005),
   http://eur-lex.europa.eu/LexUriServ/site/en/com/2005/
   com2005_0576en01.pdf
3. Rinaldi, S.M., Peerenboom, J.P., Kelly, T.K.: Identifying, understanding, and analyzing critical infrastructure interdependencies. IEEE Control Systems Magazine, 11–25 (2001)
4. G.T.I.S. Center: Emerging cyber threats report for 2009 (2008),
   http://www.gtisc.gatech.edu/pdf/CyberThreatsReport2009.pdf
5. NIST SP800-82: Draft guide to industrial control systems (ICS) security (2008),
   http://csrc.nist.gov/publications/drafts/800-82/
   draft_sp800-82-fpd.pdf
6. Chan, H., Perrig, A.: Security and privacy in sensor networks. IEEE Computer 36(10), 103–105 (2003)
7. Perrig, A., Stankovic, J., Wagner, D.: Security in wireless sensor networks. ACM Commun. 47(6), 53–57 (2004)
8. Kagan, H.: Interview about wireless devices adoption in the industry and the future trends. Frost & Sullivan (2008), http://www.teknikogviden.dk
9. Berra, J.: Emerson first to offer WirelessHART automation products (2008),
   http://www.controlglobal.com/industrynews/2008/082.html
10. Hoske, M.T., McPherson, I.: Industrial wireless implementation guide. Control Engineering (2008), http://www.controleng.com/article/CA6584939.html
11. Masica, K.: Recommended practices guide for securing ZigBee wireless networks in process control system environments, Draft (2007),
    http://csrp.inl.gov/Documents/Securing%20ZigBee%20Networks%20in%
    20Process%20Control%20Systems%20Environments.pdf
12. IEEE 802.15.4 Standard, Wireless Medium Access Control (MAC) and Physical Layer (PHY) Specifications for Low Rate Wireless Personal Area Networks (LR-WPANs)
13. Noonan, T., Archuleta, E.: The National Infrastructure Advisory Council's final report and recommendations on the insider threat to critical infrastructures (2008), http://www.dhs.gov/xlibrary/assets/niac/niac_insider_threat_to_
    critical_infrastructures_study.pdf
14. Keeney, M.: Insider threat study: Computer system sabotage in critical infrastructure sectors, Executive summary (2005),
    http://www.secretservice.gov/ntac/its_report_050516.pdf
15. Welander, P.: Securing legacy control systems (2009),
    http://www.controleng.com/article/
    307540-Securing_Legacy_Control_Systems.php
16. Wikipedia: Next generation networking, NGN all-IP (2008),
    http://en.wikipedia.org/wiki/NextGenerationNetworking
17. Fonash, P.M.: Cybersecurity & Communications (CS&C) overview, Technology trends, & challenges (2008),
    http://events.sifma.org/uploadedFiles/Events/2008/BCP/
    Fonash%20presentation.pdf

18. Kim, R.-H., Jang, J.-H., Youm, H.-Y.: An efficient IP traceback mechanism for the NGN based on IPv6 Protocol (2008),
    http://jwis2009.nsysu.edu.tw/location/paper/An%20Efficient%20IP%20Traceback%20mechanism%20for%20the%20NGN%20based%20on%20IPv6%20Protocol.pdf
19. NSTAC: Next Generation Networks Task Force, Appendices (2006),
    http://www.ncs.gov/nstac/reports/2006/NSTAC%20Next%20Generation%20Networks%20Task%20Force%20Report%20-%20Appendices.pdf
20. Zhimeng, T., Bo, W., Yinxing, W.: Security technologies for NGN (2008),
    http://www.ztebrasil.com.br/pub/endata/magazine/ztecommunications/2007year/no4/articles/200712/t20071224_162457.html
21. Yuxi, G.: IP Bearer Network for NGN (2005),
    http://wwwen.zte.com.cn/endata/magazine/ztecommunications/2005year/no3/articles/200509/t20050921_162351.html
22. DEAR-COTS project homepage (2001), http://dear-cots.di.fc.ul.pt

Adversarial Security:
Getting to the Root of the Problem*

Raphael C.-W. Phan, John N. Whitley, and David J. Parish

High Speed Networks (HSN) Research Group,
Electronic & Electrical Engineering,
Loughborough University,
LE11 3TU, UK
{R.Phan,J.N.Whitley,D.J.Parish}@lboro.ac.uk

Abstract. This paper revisits the conventional notion of security, and champions a paradigm shift in the way that security should be viewed: we argue that the fundamental notion of security should naturally be one that actively aims for the root of the security problem: the malicious (human-terminated) adversary. To that end, we propose the notion of adversarial security where non-malicious parties and the security mechanism are allowed more activeness; we discuss framework ideas based on factors affecting the (human) adversary, and motivate approaches to designing adversarial security systems. Indeed, while security research has in recent years begun to focus on human elements of the legitimate user as part of the security system's design e.g. the notion of ceremonies; our adversarial security notion approaches general security design by considering the human elements of the malicious adversary.

1 The General Security Problem

This paper sets out to revisit the conventional notion of security. In essence, conventional security represents the security advocate as a boxed-in non-initiator, in that (technical) security mechanisms therein aim to *protect* the good guy against or *cope* with, anticipated attacks. Quoting from [25], the general view is that "security is inherently about *avoiding* a negative". By design, the advocate is not equipped with the ability to initiate actions in the reverse direction towards the malicious adversary.

In that light, cryptographic techniques and network security techniques are traditionally *defensive* mechanisms in face of the malicious adversary. More precisely, confidentiality (encryption), integrity (message authentication codes, hash+signatures), authentication (signatures), non-repudiation (signatures) ensure that in the event of attacks, either data or identities are *protected from* unauthorized adversarial (read and/or write) access or at the very least that any attack is discovered by the victim parties; while intrusion detection or firewalls detect or block adversarial access.

---

* Part of this work adversarially motivated by coffee.

J. Camenisch, V. Kisimov, and M. Dubovitskaya (Eds.): iNetSec 2010, LNCS 6555, pp. 47–55, 2011.

Intrusion tolerance, network resilience and proactive cryptography [8] (including forward security [9], key insulation [14], intrusion resilience [13], leakage resilience [26,23]) techniques are of similar nature, emphasizing on being able to *cope* with or *survive* adversarial attacks.

While it must be said here that the network forensics approach does to some extent provide a channel to get back at the malicious adversary, this is via non-technical means, i.e. legal actions. Another emerging approach, non-technical as well, is that of security economics [7] that can also be seen as more proactive rather than simply defensive.

Taking a holistic view of the security problem, we would like to champion a paradigm shift in the way that security should be viewed, by arguing that the fundamental notion of security should naturally be one that *actively* aims to tackle the root of the security problem: the malicious adversary.

We also champion in this paper the fact that security should fully exploit the fact that the adversary is human-terminated; thus in terms of proactively addressing this root of the security problem, one should bear in mind that the human adversary lives in the real world and is thus influenced by real world factors aside from technical ones. Essentially, security pits human ingenuity (designer) against human ingenuity (adversary). While security research in recent years has begun to consider human factors within security designs in view that legitimate security users are often human (this makes a lot of sense since attackers have long been exploiting this weakness, e.g. social engineering), less research has concentrated on designing security by considering that adversaries are also human-initiated, although to some extent the research direction popularized by CAPTCHA [6] in considering how to identify if a human is present during web based authentication dates back to the work of Naor [22] in 1996.

2   Adversarial Security Design

We propose the notion of *adversarial security*. The adversarial angle of this notion is twofold.

First, it emphasizes on the ideal that security should be the resultant equilibrium established after fair play (to some extent) among all parties, whether honest or malicious. This is akin to the adversarial process e.g. in adversarial legal systems or adversarial politics which is game-like in nature. In contrast, the conventional notion of security does not really capture this since techniques therein are less symmetric in terms of the activeness, i.e. the malicious adversary is the active initiating party while the attacked party is the non-active defending or coping party. What is worse, the adversary bears no consequences from his/her actions nor from actions of the other non-malicious parties, while the non-adversarial parties bear the consequences of their own actions (e.g. lack of emphasis on security increases risk of being attacked) and even those of the adversary. Furthermore, although the provable security paradigm also adopts a game-like approach to defining security, it resembles less the fair play element between opposing sides that should be the nature of an adversarial process.

Second, our notion is so-called adversarial in the sense it aims to emphasize on and get to the root of security problems, i.e. the malicious human-terminated adversary.

2.1   Framework

We start by thrashing out the framework that should influence the design of adversarial security mechanisms.

The root of the problem is the adversary, and the initiator at the adversarial side is a human (indeed, in our present times unlike science fiction where AI machines can be as malicious as humans, behind every adversary is a malicious human). Designing the human adversary in, rather than leaving this out as a non-technical social engineering issue, ties in well to the concept of ceremonial design [15] that is recently gaining popularity e.g. Bluetooth's pairing and schemes based on short authenticated strings [27], where humans are included in the security mechanism design.

Including non-technical issues into the design, including human factors, is also the approach taken by the discipline of systems engineering, which provides techniques to bring different areas of expertise, including both technical and non-technical, into a cross-disciplinary approach, or set of approaches, to solve problems. Using system engineering techniques to solve security problems is likely to provide more robust solutions. Typically the sectors included within a systems engineering brief are categorised into *lines-of-development*. The lines of development are classified by the acronym *TEPID OIL* - standing for: Training; Equipment; Personnel; Infrastructure; Doctrine and concepts; Organization; Information; and Logistics [3]. Taking a systems engineering approach to security means taking into account factors that are not just technical, so would include human factors like motivation and cost.

We can holistically model the adversarial side by classifying the factors that affect this human-involved adversary into differing layers of abstraction in top-down manner:

- Top layer: Psychological (human element)
  - motivation (to mount the attack)
    * benefit (derived from the attack)
    * cost (of attack actions)
    * risk (of being held responsible)
  - social implications
    * reputation (peer status)
- Middle layer: Real World
  - physical consequence to the human adversary
  - legal implications
- Bottom layer: Technical
  - hardness (of attack actions)
  - cost (of attack actions)
  - time (taken by attack actions)

Each of the layers need to be defined as time-varying. Indeed, it is easy to see that technology advances over time, and currently infeasible technical actions may not necessarily be so in the future. Real world factors may also vary with time, e.g. changes in cyberlaws. While motivational factors do tend to be time-invariant, yet for generality the top layer can also be defined as time-varying, to capture for instance the dwindling in value of the protected data e.g. stock market information is no longer useful when it becomes obsolete; thus there is no longer any motivation to mount attacks to acquire obsoleted market info.

Most of the conventional security techniques aim to tackle the security problem by targetting the bottom abstraction layer i.e. the technical layer. For instance, provably secure cryptographic constructions rely on the assumption that it is computationally infeasible (in terms of hardness, cost and/or time) for an adversary to mount technical attack actions.

In contrast, we can use the above listed layer abstractions to define a framework that is adversarial in the sense that the non-malicious parties in their interactions with the adversary are also provided the capability to mount active responses aimed at affecting the factors within those abstraction layers. This interaction can be modelled as an adversarial game. For instance, an adversarial security notion capturing the psychological factor is as follows. Let $state_0^x$ denote the initial value assigned by the adversary to the motivation factor x, and $state_1^x$ denote the state of this at the conclusion of the adversarial game. Then the adversarial advantage can be defined as the difference between $state_0^x$ and $state_1^x$. The security mechanisms are deemed successful if $state_1^x << state_0^x$; i.e. the adversary's motivation is significantly reduced as a consequence of being involved in the attack.

Aside from influencing the design and redesign of security systems, this framework can also be used to create adversarial profiles, e.g. for network forensics and evidence construction.

2.2   Approaches

To understand this holistic, systems engineering-like approach to security systems, here we propose the following approaches be included in the overall view of solving security:

Approach 1: getting nearer to the adversarial source. We can do this in two ways.

- by abstraction: with reference to the abstraction layers, mechanisms can be designed to target factors nearest to the human adversary's mind, i.e. psychological factors. The lower down the abstraction layers, the more external and less attached it is to the adversary; e.g. if a security mechanism complicates the adversary's attack technically, then s/he can retry with another technical one. In contrast, if we design a mechanism that increases the adversary's risk of being caught for attacks s/he had mounted independent of the specific technicalities, this factor remains to internally affect the adversary's motivation. At the same time, different adversarial groups are presenting themselves, and as their motives are removed from previous adversaries, the technical implementation of their attacks are different too [4]. We suggest that involving social understanding and prevention of the motivation of an attack will lead to a stronger system.

This is also an appropriate approach when attempting to prosecute suspects of network attacks: the gathered evidences are considered not only with respect to the technical layer, e.g. packet headers, but evidences are also considered at the psychological layer e.g. the suspect's intent.

- spatially: alternatively, approach 1 could mean getting nearer to the adversary's location, e.g. rather than having firewalls essentially boxing in the attacked machine, to instead have the adversary's network provider or nearby intermediaries boxing in the adversary at his machine. This is beneficial for the network security setting. More precisely, less inconvenience is caused to the attacked machine because it is not penalized for being attacked; instead, the adversary feels the consequences of his/her attack.

There is one caveat of this approach if only a technical solution is employed. Trying to contain the adversary at the source would work for a conventional Denial of Service (DoS) attack, but is much harder to see effect if it is a Distributed Denial of Service (DDoS) attack where the adversary uses multiple distributed locations to launch the attack. In contrast, adopting approach 1 in the sense of abstraction would mean targeting a higher abstraction layer, i.e. the motivation of the adversary, and this will be equally effective to both DoS and DDoS attacks.

Approach 2: legitimate fightbacks. The idea of fighting back or striking back has been proposed in network security literature [16,30,20], albeit sparingly. Quoting [12]: "returning fire has a benefit in suppressing attack activity...a strong offense is a good defense". For instance, [16] reports on how Fortune 500 companies have admitted that they have the capability (via installed software) of counter-attcking hackers. Such measures include flooding the hacker's system. The study also reported that many companies would rather trust their own strikeback capabilities than summon the enforcement authorities. The caveat is that there may be legal issues [21], and furthermore one cannot always ascertain who the adversary is.

[28,29] suggest to fight fire with fire, i.e. combat DoS attacks by having the legitimate users also launch a kind of DoS on the network resources. This is legitimate since the attack is not directly targeted at the adversary per se, although it causes the same kind of inconvenience to legitimate users as would a DoS attack since a lot of retransmissions are involved and thus wastage of network resources. There are network protocols to help the victim require the adversary to desist, for example IPCAF [5], although an obvious problem with any such protocols is general take-up: it would require all Internet Service Providers (ISPs) to implement additional protocols within their networks.

Rather than actively 'fighting back' towards the adversary using the same weapon as the adversary, which may lead to indefinite vengeful cyber-warfare where the non-malicious parties will be no better than the adversary, we advocate instead to consider legitimate ways and fighting fire with non-fire. Essentially, this means that we use ways that are different from the ones used by the adversary; e.g. we do not launch DoS style responses to DoS attacks, or upon viral infection we do not attempt to propagate the virus back to the source. This sort of legitimate fightback is an interesting open research direction.

Some ways of legitimate fightback can be designed, with reference to the framework of abstraction layers:

- de-motivation: we can design a system such that the benefit derived from an attack is much less than the attack cost. For instance, the business model used by Apple where third party developers build applications that sell for a song; malicious users are then no longer motivated to copy nor pirate these applications since the cost of doing so is not significantly less than actually buying the applications legitimately.
- social pressure: humans tend to bow down to peer pressure. As an analogy: rather than directly punish a misbehaving pupil, a class teacher could subtly get the class to frown on (e.g. laugh at) the pupil by making a cynical remark. In similar vein, the social networking site Facebook.com retracted its new policy to retain user data online after public protest [17].
- removal of cheap resource: distributed attacks require a number of slave hosts in the network, commonly the owners of these machines would not intend their machines to be part of a distributed attack. The implementation of ingress firewalls to prevent attacks on a system is well understood and well used in many Internet sectors [1]. The implementation of ingress firewalls as part of a domestic and commercial Internet connection is now thought of as normal and expected. This has prevented many simple virus attacks on otherwise vulnerable machines. We suggest that egress firewall filtering, if implemented on the scale that ingress firewall filtering has been, will have a similar tide-changing effect in reducing adversarial network traffic.

Approach 3: human involvement. Keeping in mind our approach to design the human element of the adversary right into a security system, the human involvement approach advocates the following design strategies:

- human-tractable but machine-intractable tasks: the strategy here is to have security-critical actions require human mediations; this means the human adversary cannot fully automate his/her attack steps since the tasks cannot be easily performed by machines, and therefore affects the cost and time subfactors as well as increases attack consequences on the adversary. Examples of security systems that use this approach are systems based on short authenticated strings, for example Bluetooth, which requires human-communication channels; and CAPTCHA to combat machine-automated spamming or DoS attacks. While the approach is nice, particular techniques to implement such an approach should be designed with care. For instance, it was recently shown that typical CAPTCHA techniques are known to be weak [2].
- human-involved ceremonies: the strategy of designing systems to include the human terminals, so-called ceremonies [15,18,19] allows to capture not just

the technical issues but the human elements of users of such systems as well, and therefore relevant issues such as social engineering no longer need to be regarded as out of scope during the design stage.

Approach 4: every action has a reaction. This strategy is to design a system where every action by a party (irrespective of whether it is an honest party or a malicious one) causes a reaction i.e. leads to an effect on the party itself, e.g. each action costing monetarily or resource-wise. For instance, in some security systems for wireless networks [31], incentive based schemes are designed where credit is globally distributed to all parties during setup or on enrollment into the system, and where credits are awarded to or deducted from a party based on its actions.

In this setting, malicious parties will be penalized directly from their attack actions, even if an attack attempt is not successful. This is also particularly relevant in the network security scene, e.g. DoS attacks, spams or spits, where each generated traffic towards the target victim machine incurs a reaction back to the adversary. What is more, non-attacking parties who wrongly accuse others will want to think carefully because their accusing actions will also cause a reaction e.g. cost.

Approach 5: being stateful and bearing grudges. Related to approach 4, in the sense that each attack attempt should update the system state so that there is a reaction back to the adversary that affects the adversary's subsequent actions, the approach here tackles attacks of the exhaustive type, e.g. brute force password (or secrets of low entropy) guessing attacks. The gist is that the security system should remember each attack attempt (even if the attempt does not lead to a successful attack), and be stateful such that the subsequent attack attempt would require significantly more adversarial effort to mount. This idea is used by Bluetooth [10] to discourage bruteforce guessing, by having each repeated attempt lead to a waiting time that is exponentially increasing. On a related note, the idea of bearing grudges has also been applied to discourage misbehaviour by selfish parties rather than discourage attacks, e.g. [11].

3 Concluding Remarks

We have advocated a paradigm shift in the way we address the security problem, i.e. taking a holistics systems engineering-like approach and in doing so including an additional focus on the adversarial angle. Adversarial angle in the sense of fair play between the adversarial and attacked sides, and in the sense of getting right to the source of the problem i.e. the human-terminated adversary. We proposed to treat the factors affecting an adversary as time-varying layers of abstraction; and discussed five approaches with this in mind.

Security should not remain as a purely defensive strategy, quoting [30]: "sitting back and waiting for attackers is a strategy doomed to failure... defensive wars are not winnable".

Acknowledgement

We thank the iNetSec 2010 anonymous reviewers and non-anonymous attendees for comments and interest in this research direction.

References

1. Chapman, D.B., Zwicky, E.D., Russell, D.: Building Internet Firewalls. O'Reilly & Associates, Inc., Sebastopol (1995)
2. Hernandez-Castro, C.J., Ribagorda, A.: Remotely telling humans and computers apart: An unsolved problem. In: Camenisch, J., Kesdogan, D. (eds.) iNetSec 2009. IFIP Advances in Information and Communication Technology, vol. 309, pp. 9–26. Springer, Heidelberg (2009)
3. Kerr, C., Phaal, R., Probert, D.: A Framework for Strategic Military Capabilities in Defense Transformation. In: International Command and Control Research and Technology Symposium (2006)
4. BBC News. Political Hacktivists Turn To Web Attacks (2010), http://news.bbc.co.uk/1/hi/technology/8506698.stm; This is an electronic document. Date of publication: February 10, 2010. Date retrieved: February 10, 2010. Date last modified: February 10, 2010
5. Wu, C.-H., Huang, C.-C.A., Irwin, J.D.: Using Identity-Based Privacy-Protected Access Control Filter (IPACF) to Against Denial Of Service Attacks and Protect User Privacy. In: Proc. SpringSim 2007, San Diego, CA, USA, pp. 362–369 (2007)
6. von Ahn, L., Blum, M., Hopper, N.J., Langford, J.: CAPTCHA: Using Hard AI Problems for Security. In: Biham, E. (ed.) EUROCRYPT 2003. LNCS, vol. 2656, pp. 294–311. Springer, Heidelberg (2003)
7. Anderson, R., Moore, T.: The Economics of Information Security. Science 314(5799), 610–613 (2006)
8. Barak, B., Herzberg, A., Naor, D., Shai, E.: The Proactive Security Toolkit and Applications. In: Proc. ACM CCS 1999, pp. 18–27 (1999)
9. Bellare, M., Miner, S.: A forward-secure digital signature scheme. In: Wiener, M. (ed.) CRYPTO 1999. LNCS, vol. 1666, pp. 431–448. Springer, Heidelberg (1999)
10. Bluetooth SIG, Bluetooth Core Specifications v4.0 (December 17, 2009)
11. Buchegger, S., Le Boudec, J.Y.: Nodes Bearing Grudges: Towards Routing Security, Fairness, and Robustness in Mobile Ad Hoc Networks. In: Proc. PDP 2002, pp. 403–410 (2002)
12. Cohen, F.: Managing Network Security: Returning Fire. Network Security 1999(2), 11–15 (1999)
13. Dodis, Y., Franklin, M.K., Katz, J., Yung, M.: Intrusion-resilient public-key encryption. In: Joye, M. (ed.) CT-RSA 2003. LNCS, vol. 2612, pp. 19–32. Springer, Heidelberg (2003)
14. Dodis, Y., Katz, J., Xu, S., Yung, M.: Key-insulated public key cryptosystems. In: Knudsen, L.R. (ed.) EUROCRYPT 2002. LNCS, vol. 2332, pp. 65–82. Springer, Heidelberg (2002)
15. Ellison, C.: UPnP Security Ceremonies: Design Document (October 2003), http://www.upnp.org/download/standardizeddcps/UPnPSecurityCeremonies_1_0secure.pdf
16. Gengler, B.: Strikeback. Computer Fraud & Security 1999(1), 8–9 (1999)

17. Johnson, B., Hirsch, A.: Facebook Backtracks after Online Privacy Protest (February 19, 2009), http://Guardian.co.uk
18. Karlof, C., Tygar, J.D., Wagner, D.: Conditioned-safe Ceremonies and a User Study of an Application to Web Authentication. In: Proc. NDSS 2009 (2009)
19. Karlof, C., Tygar, J.D., Wagner, D.: Conditioned-safe Ceremonies and a User Study of an Application to Web Authentication. In: Proc. SOUPS 2009 (2009)
20. Jayawal, V., Yurcik, W., Doss, D.: Internet Hack Back: Counter Attacks as Self-Defense or Vigilantism? In: Proc. ISTAS 2002 (2002)
21. Matsuura, J.H.: "Digital Victim or "Vigilante": Legal and Ethical Limits to Online Self-Defense. In: Proc. Ethicomp 2004, pp. 629–634 (2004)
22. Naor, M.: Verification of a Human in the Loop, or Identification via the Turing Test (September 1996),
    http://www.wisdom.weizmann.ac.il/~naor/PAPERS/human_abs.html
23. Phan, R.C.-W., Choo, K.-K.R., Heng, S.-H.: Security of a leakage-resilient protocol for key establishment and mutual authentication. In: Susilo, W., Liu, J.K., Mu, Y. (eds.) ProvSec 2007. LNCS, vol. 4784, pp. 169–177. Springer, Heidelberg (2007)
24. Schneier, B.: The Psychology of Security. Communications of the ACM 50(5), 128 (2007)
25. Schneier, B.: How the Human Brain Buys Security. IEEE Security & Privacy 6(4), 80 (2008)
26. Shin, S., Kobara, K., Imai, H.: Leakage-resilient authenticated key establishment protocols. In: Laih, C.-S. (ed.) ASIACRYPT 2003. LNCS, vol. 2894, pp. 155–172. Springer, Heidelberg (2003)
27. Vaudenay, S.: Secure communications over insecure channels based on short authenticated strings. In: Shoup, V. (ed.) CRYPTO 2005. LNCS, vol. 3621, pp. 309–326. Springer, Heidelberg (2005)
28. Walfish, M., Balakrishnan, H., Karger, D., Shenker, S.: DoS: Fighting Fire with Fire. In: Proc. HotNets 2005 (2005)
29. Walfish, M., Vutukuru, M., Balakrishnan, H., Karger, D., Shenker, S.: DDoS Defense by Offense. ACM SIGCOMM Computer Communication Review 36(4), 303–314 (2006)
30. Welch, D.J., Buchheit, N., Ruocco, A.: Strike Back: Offensive Actions in Information Warfare. In: Proc. NSPW 1999, pp. 47–52 (1999)
31. Zhang, Y., Lou, W., Fang, Y.: SIP: a Secure Incentive Protocol against Selfishness in Mobile Ad Hoc Networks. In: Proc. IEEE WCNC 2004, pp. 1679–1684 (2004)

Practical Experiences with Purenet, a Self-Learning Malware Prevention System

Alapan Arnab[1], Tobias Martin[2], and Andrew Hutchison[1]

[1] T-Systems South Africa, International Business Gateway, New Road
Midrand, 1685, South Africa
{alapan.arnab,andrew.hutchison}@t-systems.co.za
[2] Deutsche Telekom Laboratories, Deutsche Telekom Allee 7
64295, Darmstadt, Germany
tobias.martin@telekom.de

Abstract. This paper introduces Purenet, which is a self-learning malware detection system aimed at avoiding zero-day attacks and other delays in patching application systems when attacks are identified. The concept and architecture of Purenet are described, specifically positioning anomaly detection as the system enabler. Deployment of the system in an operational environment is discussed, and associated recommendations and findings are presented based on this. Findings from the prototype include various considerations which should influence the design of such security software including latency considerations, multi protocol support, cloud anti-malware integration, resource requirement issues, reporting, base platform hardening and SIEM integration.

1 Introduction

Malware (viruses, worms, spyware etc) is often cited as one of the main electronic security problems [3],[4]. While active attacks against a system are often targeted, and usually require a high degree of knowledge (of both technology and the targeted system), malware attacks are generic, high volume and - with the advent of relevant tools - can be largely automated.

Systems to prevent malware prevention therefore have a high visibility, and have more awareness in the general public than more advanced security solutions such as hardware encryption modules. Despite the larger awareness of such solutions, there are two problems with the effectiveness of such solutions:

1. Anti-Malware (and in general, endpoint security) solutions are not always kept up-to date. [9]
2. Systems (including all the relevant software on the system) are not patched with regularity, allowing certain types of malware to spread and infect before they can be detected by anti-malware solutions.

Furthermore, anti-malware solutions will often struggle to identify new threats that have not yet been identified (such as zero-day attacks, or attacks launched

J. Camenisch, V. Kisimov, and M. Dubovitskaya (Eds.): iNetSec 2010, LNCS 6555, pp. 56–69, 2011.

on a software vulnerability that has no patch). The rapid growth of the Internet has also increased the number of malware threats, with a trend in faster exploitation of unremediated vulnerabilities. [10]

The Netcentric Security project, conducted at Deutsche Telekom Laboratories, considered a two pronged approach - firstly to prevent known malware from entering a protected network domain (in similar fashion to an Intrusion Prevention System); and secondly to detect new malware using machine learning techniques by Purenet. After a new malware has been detected, a signature of the file containing the malware is derived automatically and the intrusion prevention device is updated.

The structure of the paper is as follows: background information to motivate the approach is provided, Purenet is introduced as a concept, and then described in terms of its approach and architecture. The testing of the system is then outlined, and recommendations and findings based on live deployment of a system prototype are provided. Future work and conclusions are then provided.

2   Background

The approach of attacking malware dissemination at the Internet Gateway, is increasingly attracting attention - and analysts such as Gartner recommend increasing focus on this approach [2]. While Gartner's recommendations are based on leveraging existing technologies, such as Intrusion Prevention Systems and e-mail/URL filtering applications, Purenet's approach is to further increase the effectiveness of such systems by *proactively* scanning for new malware.

Figure 1 provides insight into why a proactive security approach is desirable. When unknown or new malware starts to circulate, a window of opportunity arises between the identification of such malware and new anti-virus patches and/or signatures being released. Attackers aim to exploit the Window of Opportunity, the time AV and software vendors need to identity new threats and provide patches. New malware can spread until signature databases are updated. Security flaws in software may be exploited until patches are available.

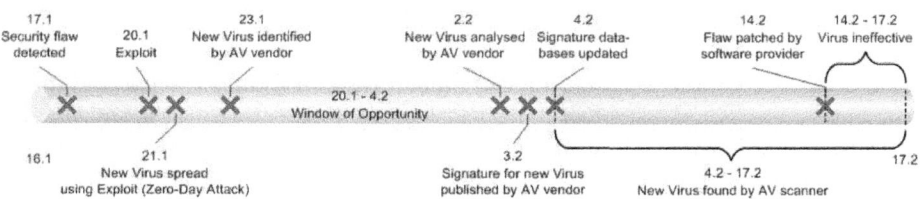

Fig. 1. Zero Day Attacks

As detailed in figure 1 a security flaw arises in a timeline starting on 17th January, and a window of opportunity exists between 20th January (when an exploit is created / released) and the 4th February when signature updates are released. So called 'zero day attacks' can take advantage of such windows of opportunity to do large amounts of damage in a short time.

3   Purenet as Concept

The "Netcentric Security" project, at Deutsche Telekom Laboratories, has a two pronged approach:

1. to prevent known malware from entering a protected network domain (similar to an Intrusion Prevention System); and
2. to detect new malware using machine learning techniques by Purenet.

In essence, the Purenet goal can be stated as detection of unknown malware (viruses, trojans or spyware) by code classification using machine learning techniques used in Artificial Intelligence (AI).

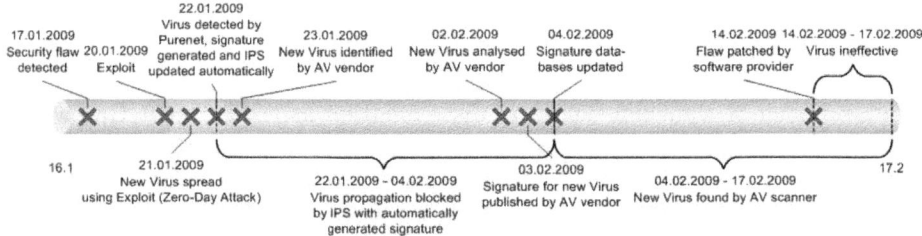

Fig. 2. Detection of unknown Malware

In contrast to Figure 1, Figure 2 shows how a system based on proactive malware detection can stop virus propagation immediately. New threats are detected by an AI system, IPS signatures are generated automatically. Identified threats are stopped at the entry to the network until AV signatures or software patches are available. The window of opportunity for attacks, and zero day possibility, are eliminated if direct identification of threats is possible.

Figure 3 illustrates the positioning of Purenet within an Internet services environment. Purenet is deployed in a DMZ and must be connected to the copy port of an IDS/IPS or network tap. Purenet performs sniffing and file reconstruction, new threat detection in files, signature generation and filter device update and alerting.

A data link between the Purenet system and other Anti-Virus (AV) or Intrusion Prevention Systems (IPS) is configured so that any findings of the Purenet system can be propagated to other scanning devices immediately. The concept of Purenet is to do analysis and update in parallel to the inspection by the IPS with a slight time delay.

From a scalability point of view a network of distributed Purenet instances can also be deployed, with findings at any one instance propagated to other Purenet deployments for the purpose of updating associated AV or IPS devices at their locations.

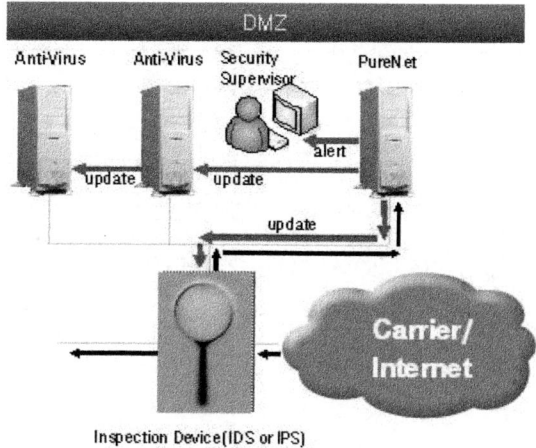

Fig. 3. Purenet deployment for an Internet Service Provider

4   Purenet Approach and Architecture

Purenet was designed with a modular architecture to allow live scanning of Internet data, across multiple file types for known malware. Identification of unknown malware is facilitated by time-delayed analysis of files in a so-called New eThreat Detection Module (NeDM). Once this NeDM classifies a file as malware, or as containing malicious code, a signature is generated automatically and directly so that the same pattern can be detected subsequently by the IPS-like live scanning process. In the initial implementation only HTTP traffic was monitored, and only Microsoft Windows executables and DLL files were scanned for malware.

Figure 4 shows the conceptual architecture of Purenet. Known malware is identified and blocked by the IDS/IPS appliance. Remaining traffic is monitored and collected by the Data Stream Manager (DSM). The New eThreat Detection Module (NeDM) analyzes the captured files and the Signature Builder (SB) is activated in order to synthesize a signature of newly detected malware. The Storage Manager (SM) stores hash values for all analyzed files to detect recurrences. Purenet Control Center may be used by security experts to resolve conflicts.

4.1   Data Stream Manager (DSM)

The DSM extracts packets from a TCP/IP stream and extracts relevant files for the New eThreat Detection Module. It is fed with IP packets from the network and performs as a sniffer, filtering packets to discard traffic which is not relevant for the NeDM.

IP traffic is first filtered by the packet filter passing only http (TCP/IP port 80) and snf SMTP (TCP/IP port 25) packets to the capture module. The capture

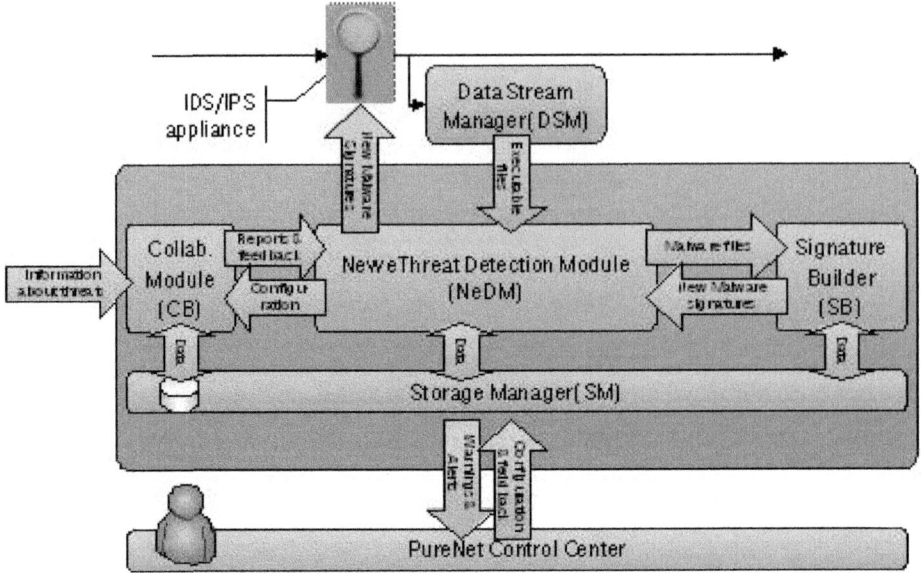

Fig. 4. Conceptual architecture of Purenet

module delivers captured packets to the session extractor module reassembling a session and passing on to the file extractor module. Whenever a file is extracted, it is filtered by the recurrence filter to reduce duplications of files to be inspected by NeDM. New files are stored in a local file buffer before they are inserted in the Storage Manager and passed to the New eThreat Detection Module. The local file buffer is also used by the recurrence filter to check for duplications.

The Purenet prototype only contains a file extractor plug in for files contained in http and SMTP sessions. However, since the DSM has a modular architecture, additional plug ins, e.g. for POP3, IMAP4 etc. might be added.

4.2   New eThreat Detection Module (NeDM)

The NeDM is the most important module in the Purenet architecture. It does not process information in real-time. The architecture of NeDM is depicted in Figure 5. New files are received from the DSM and analyzed by various plug-ins which employ the feature extractor component that extracts for each plug-in the necessary features for the analysis process. Results from plug-ins are forwarded to the risk weighting module generating a final recommendation by combining the results from the plug-ins.

The Unknown Malware processor is controlling the whole detection process. Its input is files collected by the DSM which are inserted into the Storage Manager (SM). If the file already exists, NeDM updates its statistics in the SM. Otherwise it is stores in the SM and its status is set tu 'unchecked'. Whenever NeDM is idle it retrieves the unchecked file with highest priority.

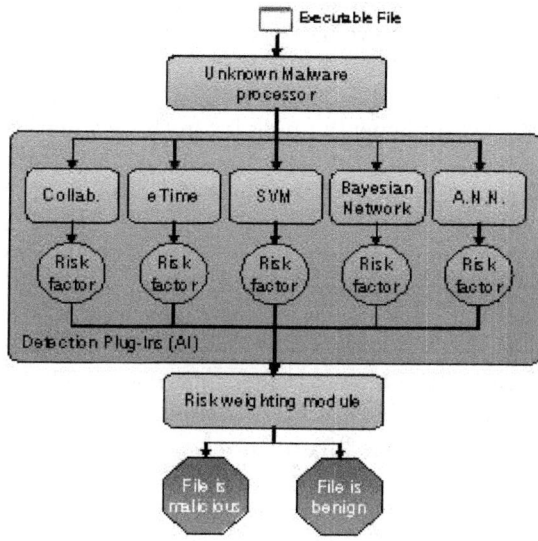

Fig. 5. Conceptual architecture of NeDM

The various detection plug-ins analyze the file using different algorithms. Then, the results are combined by the risk weighting module, which takes also alerts from the Collaborative Module into account. If the risk grade is above a maximum threshold, the file is classified as malicious and a signature is extracted. If the malicious risk grade is below a minimum threshold, the file is classified as benign. Otherwise it is forwarded to the Purenet Control Center for further inspection by human experts.

4.3 Detection Plug-ins

To enable flexibility in employing various algorithms for identifying new malware, plug-ins are used. All plug-ins have a similar interface, a file as input and a risk factor as output. Generally NeDM supports two types of plug-ins:

1. Static (structural) analysis plug-ins: An executable file is analyzed before being executed. Structural features are extracted from each file and analyzed by various machine learning techniques, such as Decision Trees, Bayesian Networks, etc.
2. Behavioral (dynamic) analysis plug-ins: Analyze executable files by executing the program in a supervised run-time environment.

Dynamic plug-ins require longer time to reach a decision as the file needs to be executed and it might take a long time until it starts its malicious activity. On the other hand, static plug-ins generate a decision in a few seconds.

Dynamic plug-ins should hence be used selectively and not each time NeDM inspects a new file. They might be used for instance on files that were classified as "expert review".

In the evaluation of Purenet presented in this paper static plug-ins based on machine learning algorithms were used exclusively.

These plug-ins work as follows: First, features that represent the file are extracted. Then the plug-in applies the classification model on the features and returns the results. The implemented plug-ins in this evaluation employ machine learning classifiers which were trained using the WEKA software (Witten and Frank, 2005), commonly used for these types of tasks [11].

In the evaluation nine different classifiers were used: Bayesian Networks, Nave-Bayes, K-Nearest Neighbors, Hyper Pipes, VFI, J48 Decision Tree, Random Forrest, OneR and PART. More details can be found in [8] and [1], see also [5],[6] and [7].

In the training phase a repository of 7,688 malicious files and 22,735 benign programs running on a Windows XP machine were used. For each file, three types for features were extracted: n-grams based features, Portable Executable (PE) features and function-based features.

The n-grams (3-grams, 4-grams, 5-grams and 6-grams of bytes) were extracted from the binary representation of a file. The PE headers of Win32 binaries (EXE or DLL) might indicate that a file is infected by a virus, e.g. inconsistencies between different parts of version numbers or internal/external name of a file. In total 88 features are extracted from PE headers. The Function method is a new method for extracting features from files: The beginning and end-points of functions in binary code is marked. Using the marks the following 17 features are extracted:

- Size of file; File's entropy value;
- Total number of detected functions;
- Average size and size of the longest and shortest detected function;
- Standard deviation of size of detected functions;
- Number of functions divided into fuzzy groups by length in bytes;
- Function ratio and function code ratio.

4.4   Risk Weighting

The risk weighting module provides a final rank for each file suspected as malware. It collects all risk factors from the relevant plug-ins and calculates a weighted rank. If the final rank is beyond a threshold, the file will be transferred to the Signature Builder which will construct a unique signature.

For the Purenet test the Distribution Summation weighting process is used, which is quite simply and outperformed most of other weighting algorithms. It accumulates the risk factors generated by the plug-ins for each class (i.e. benign and malicious). The class with the highest summed grade is chosen.

5   Purenet Testbed

A live testbed of the solution was implemented in the Internet gateway of T-Systems South Africa, which serves only large corporate clients. The intention of the testbed was to investigate the feasibility of the solution for corporate customers and to investigate the effectiveness of the detection module in a large environment with many concurrent connections. While the Purenet solution did not change the operational environment itself, all data collected and analysed was real world data, using a real time traffic feed.

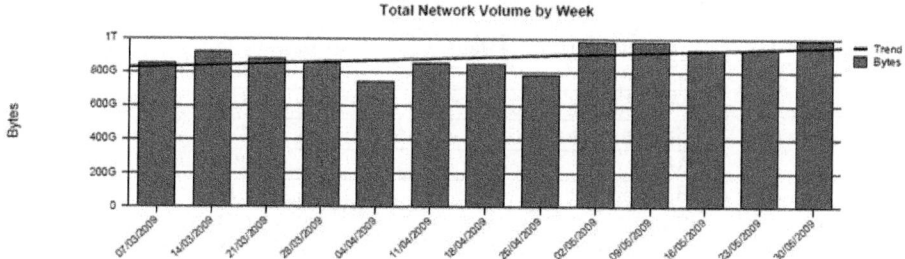

Fig. 6. Total Network Volume, per week (T-Systems SA Internet Gateway) in Bytes

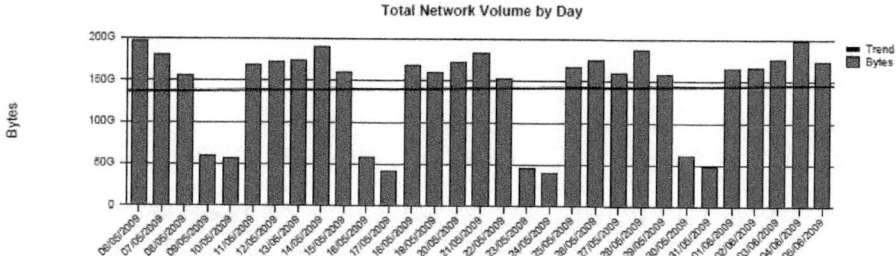

Fig. 7. Total Network Volume, per day for May 2009 (T-Systems SA Internet Gateway) in Bytes

Figures 6 and 7 provide typical traffic volumes through the T-Systems South Africa Internet gateway. It was in this context that the Purenet prototype was tested.

In addition to the actual throughput of data transversing the system, another important factor to consider is the number of simultaneous connections transversing the solution; as well as the traffic types of these connections. The performance of Purenet is not constrained by the data throughput but rather by the number of simultaneous connections that needs to be processed. Table 1 provides the total number of simultaneous TCP connections passing the testbed environment over a 5 minute interval at 12h50 on Friday 19 June 2009. This time period usually features the peak number of Internet users, and thus provides a good overview of the peak constraints for the testbed.

The architecture in which the Purenet prototype was deployed is shown in Figure 8. A domestic MPLS network acts as a distribution platform to/from the corporate customer community. Demilitarized zones are traversed as traffic moves between the Internet gateway and servers or users.

Table 1. Total TCP Connections over 5 Minute Interval

Application	Total TCP Connections
FTP	138
VPN	416
SSL	428
MSN	491
Skype	615
SMTP	15 681
HTTP/HTTPS	183 355

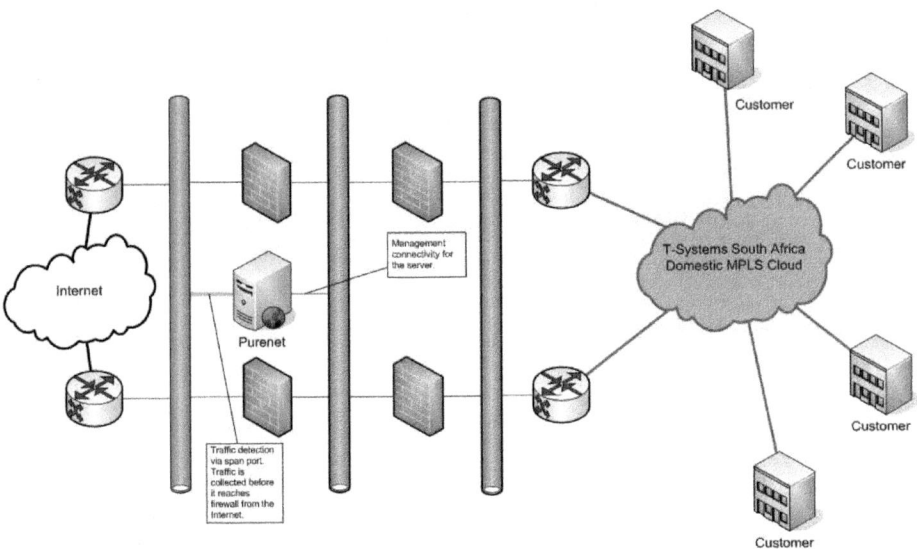

Fig. 8. Deployed Architecture for Purenet Testbed

The configuration of Purenet for the testbed was such that various detection modules were incorporated. A "Known eThreat Handling Module" was present, together with a "Data Stream Manager", "New eThreat Detection Module", "Signature Builder" and "Storage Module". A control centre was also utilized to co-ordinate this combination of known and unknown, static and dynamic detection modules. A graphical overview of this configuration is depicted in Figure 9.

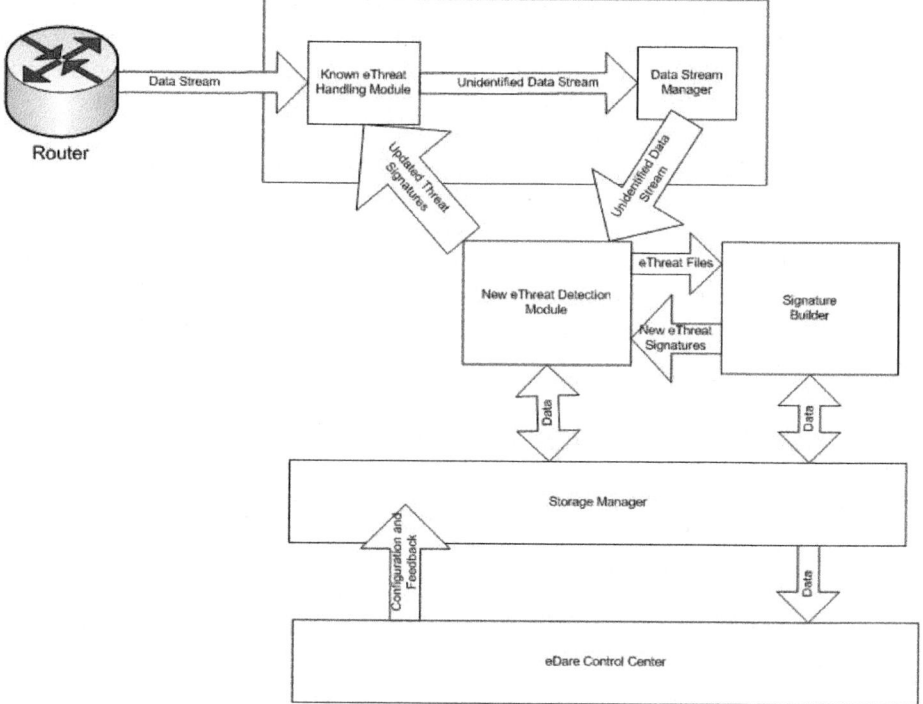

Fig. 9. Purenet Modules and Interaction (as setup on the Testbed)

6 Findings and Recommendations

The majority of findings relate directly to features and functionality requirements relevant to product development. In this section some details of each finding are presented.

6.1 Latency Impact

The practicality of the solution will be largely impacted by the speed at which it will process information; and therefore, the latency impact upon network traffic.

This will also impact the implementation design of the solution - whether it should be placed in line with the network (thus forming part of an IPS device for example) or on a span port as discussed in this paper.

The other impact of processing latency, specifically in the out-of-line deployment scenario, is that in the case of a zero-day attack, the IPS signatures will not be updated before the first zero-day attack penetrates the network (since the malware packet will most likely enter the network before Purenet concludes that the examined packet was indeed malware). While the spread of zero-day attack malware can still be largely mitigated in this approach, a total protection against detected zero-day attacks cannot be enforced; limiting the applicability of this solution. However, since an alert will be issued by the Purenet control center and a signature will be created by the Signature Builder, the zero-day attack can still be stopped even after it entered the protected network.

The in-line application of the solution will address this problem (by allowing only "clean" files through. This approach will however have an obvious latency impact, and this latency requires further analysis. Furthermore, the latency impact will differ in terms of end user experience depending on the protocol used. For example, in non-synchronous applications such as e-mail (SMTP) there is no impact on the user experience despite the fact that some e-mails are delayed (e-mails containing new files are received not before the sending MTA starts the second try).

6.2   Analysis of Malware over Multiple Protocols

The current Purenet solution only analyses http and SMTP traffic. While http and SMTP do comprise a substantial portion of Internet traffic, other protocols such as FTP, Skype and Bittorrent are also widely used and are more frequently used for transferring executable files. Support for a wider range of protocols is thus necessary for the solution to be more generally effective for its purpose.

6.3   Support for More File Types and Platforms

The current Purenet solution focuses on Microsoft Windows executables and Microsoft Windows DLL files. A significant proportion of malware exists in other file formats such as embedded files in Microsoft Office documents and Active-X applets. Detecting and mitigating vulnerabilities in Web-2 application frameworks (such as Microsoft's Silverlight and Adobe's AIR) will also enhance the value proposition of Purenet.

Another enhancement of Purenet's value proposition would be to detect malware on multiple platforms; specifically Unix based platforms such as GNU-Linux and Solaris. While there are a number of anti-malware solutions for Microsoft Windows platforms, there is little variety in Unix based anti-malware solutions. Despite the importance of Unix servers in enterprises, and the growth of Unix based operating systems such as GNU-Linux and Apple's OS X on the desktop, anti-malware solutions are rarely deployed on these platforms.

Implementing the anti-malware solution on POSIX standards, the complexity of catering for multiple Unix variants could be diminished.

6.4 Integration with Cloud Anti-Malware Solutions

In the current Purenet solution, a detection scan can produce three possible outcomes:

1. Determine that the file is not malware
2. Determine that the file is malware
3. Request that the file is examined by an expert to determine its status

In the last scenario, Purenet could utilise the emerging Cloud Anti-Malware solutions to increase the speed of resolving the status of undetermined files.

6.5 Reporting

The advantage of solutions such as Purenet, lies in the ability of the solution to generate data regarding the emergence of new malware threats; as well as adaptive solutions to combat such threats. The analysis of the data collected through the operation of solutions such as Purenet would further enhance the general understanding of malware threats; especially in environments with large diverse endpoints - such as Internet service providers.

Statistical data to back-up the functionality of Purenet is also required, and can be a separate module, directly interfaced with the database. The following reports would be expected from such a system:

- Analysis of file types over a specific time period
- Trend analysis of detected malware
- Analysis of files whose status could not be detected
- Analysis of top sources/domains of malware
- Analysis of malware based on protocol usage
- Traffic analysis in terms of clean vs malware infected files.

The above information would not only enhance the security of the networked environment, but also allow Internet service providers to better understand the traffic patterns of malware distribution and if applied correctly; top source and destination of malware distribution.

7 Future of Purenet

The Netcentric Security project has been used to introduce the concept to potential business customers of T-Systems in Germany. The data collected from Purenet showed promising results in detecting new and unknown malware.

From market analysis, feedback from assessment reports indicates that the product concept is perspicuous. However, while enterprise customers were interested in principle they would rather expect to purchase such a product from an established vendor of anti virus solutions, since there is an expectation of the product being maintained and further developed. This ultimately led to a decision to sell patents and technology to an established anti-malware vendor. There is continued development however on the analysis approach, and there are still opportunities to deploy this approach as part of a differentiated value added service for service providers.

8 Conclusion

In summary, the Purenet testbed provided a valuable confirmation of the requirement and solution approach. In terms of deploying the system in practice, the research endeavor was an example of a successful collaboration between the DT research entity and its enterprise provider unit, in terms of developing and field-testing the system. It was beneficial to deploy the Purenet solution as a testbed within the TSSA Internet gateway.

From the testbed it was possible to obtain live data from a production system, and this gave insights into areas including:

- Buffer design
- Enhancements for higher traffic volume
- Data obtained in terms of scalability of solution
- Hardware utilisation showed over-spec of systems requirements

The concepts of the Purenet testbed can be included in future projects and solutions, and the findings and recommendations are of general usefulness for internal successors and for other security researchers to consider.

References

1. Elovici, Y., Shabtai, A., Moskovitch, R., Tahan, G., Glezer, C.: Applying machine learning techniques for detection of malicious code in network traffic. In: Hertzberg, J., Beetz, M., Englert, R. (eds.) KI 2007. LNCS (LNAI), vol. 4667, pp. 44–50. Springer, Heidelberg (2007)
2. Firstbrook, P.: Why Malware Filtering Is Necessary in the Web Gateway. Published 2008-08-26 Gartner. Gartner ID: G001584595
3. Heidari, M.: Malicious Codes in Depth (2004), http://www.securitydocs.com/pdf/2742.pdf
4. Kienzle, D.M., Elder, M.C.: Internet WORMS: Past, Present, and Future: Recent worms: a survey and trends. In: ACM Workshop on Rapid Malcode, WORM 2003 (2003)
5. Moskovitch, R., Stopel, D., Feher, C., Nissim, N., Elovici, Y.: Unknown Malcode Detection via Text Categorization and the Imbalance Problem. In: IEEE Intelligence and Security Informatics, Taiwan (2008)
6. Moskovitch, R., Feher, C., Elovici, Y.: Unknown Malcode Detection - A Chronological Evaluation. In: IEEE Intelligence and Security Informatics, Taiwan (2008)
7. Moskovitch, R., Elovici, Y.: Unknown Malicious Code Detection - Practical Issues. In: 7th European Conference on Warfare and Security, Plymouth, UK (2008)

8. Moskovitch, R., Nissim, N., Elovici, Y.: Acquisition of Malicious Code Using Active Learning. In: Bonchi, F., Ferrari, E., Jiang, W., Malin, B. (eds.) PinKDD 2008. LNCS, vol. 5456, pp. 74–91. Springer, Heidelberg (2009)
9. NCSA Study (2005),
   http://www.staysafeonline.info/pdf/safety_study_2005.pdf
10. Symantec: Security Report (2006), http://www.symantec.com
11. Weka software, http://www.cs.waikato.ac.nz/ml/weka/

A Biometrics-Based Solution to Combat SIM Swap Fraud

Louis Jordaan and Basie von Solms

University of Johannesburg, Academy for Information Technology,
Cnr Kingsway and University Road, Auckland Park,
2006 Johannesburg, Republic of South Africa
louis.jordaan@gmail.com
basievs@uj.ac.za

Abstract. Cybercriminals are constantly prowling the depths of cyberspace in search of victims to attack. The motives for their attacks vary: some cybercriminals deface government websites to make political statements; others spread malicious software to do large-scale harm; and others still are monetary motivated. In this paper we will concentrate on "cyber fraudsters". At the time of this writing, a prime hunting ground for fraudsters is online banking. Millions of people worldwide use online banking to quickly and conveniently do their regular bank-related transactions. Unfortunately, this convenience comes at a price. By doing their banking online, they are vulnerable to falling prey to fraud scams such as SIM swap fraud. This paper explains what SIM swap fraud is and how it works. We will analyze the online banking payment transaction process to discover what vulnerabilities fraudsters exploit via SIM swap fraud, and then introduce a computer-based security system which has been developed to help combat it.

Keywords: Cybercriminals, Fraud, Internet Banking, Online Banking, SIM Swap Fraud, OTP, Biometrics, BIO-Swap.

1   Online Banking

Let's face it, the Information Age we live in today is significantly more fast-paced and demanding than a decade or two ago. There is no longer enough time in a day for people to finish all their work, chores and obligations. Spare time is a luxury; we do not want to spend it waiting in lengthy bank queues to do our banking transactions. We would much rather prefer to do our banking from the comfort of our home, office, or whilst enjoying a beverage at our favourite coffee shop.

Enter "online banking" (also known as Internet banking). Online banking is many banks' way of harnessing the power of the Internet to provide electronic banking services to their patrons. It allows people to remotely carry out most (if not all) of their regular banking transactions, 24 hours a day, 7 days a week, from anywhere in the world an Internet connection is available. They simply

J. Camenisch, V. Kisimov, and M. Dubovitskaya (Eds.): iNetSec 2010, LNCS 6555, pp. 70–87, 2011.

have to open their browser, navigate to their bank's online banking website, select the appropriate transaction from a menu, and then follow the instructions on screen.

The ease-of-use, speed and convenience of online banking have lead to its wide-spread adoption by millions of people around the world. According to the 2009 Consumer Billing and Payment Trends Survey [1], in the United States 4 out of 5 (or 69.7 million) households with Internet access make use of online banking services. In the United Kingdom more than 50% of Internet-using adults now bank online [2] while in the less developed South Africa where a mere 10.5% of the population has Internet access [3], 75% of those fortunate people make use of online banking [4].

2   Online Banking Fraud

In the brick-and-mortar world banks can use thick concrete walls, strong safe doors, state-of-the-art alarm systems and armed guards to protect their money from the likes of Jesse James [5] and Butch Cassidy [6]. In cyberspace however, bank robbers do not attempt to break into a safe, or storm into a bank pointing guns at people and order the cashiers to hand over the money. No, they are faceless criminals who do not have to step foot into a bank to nick other people's money. Thanks to power of the Internet, they can do so remotely from the safety of their hideaway.

Banks therefore have to rely on firewalls and intrusion detection systems (IDS) to protect their Internet-connected computers and databases (where their clients' financial information is stored) against unauthorized access and tampering. These security measures work well to keep intruders out of a bank's computer network infrastructure, but the protection they offer is unfortunately limited to the nodes within the infrastructure's perimeter. By providing online banking services to its clients, a bank opens a new window of opportunity for fraudsters: the people who make use of the online banking services.

The level of computer literacy and information security awareness varies between Internet users. Many people are under the impression that everything on the Internet can be trusted (e.g. an e-mail from a claimed source has to be legitimate, right?) and are therefore rather clueless when it comes to protecting their personal information in cyberspace. This makes them easy targets for cyber fraudsters who are devising ever more sophisticated and cunning scams to defraud unsuspecting victims out of their hard-earned savings.

Business is booming for cyber fraudsters as recent studies (at the time of this writing) have shown that online banking fraud is rising rapidly. According to the Fraud the Facts 2009 report [7] that was published by APACS (the UK Payments Association), the losses to online fraud scams such as phishing and spyware during 2008 totalled a whopping £52.5 million in the UK alone — An increase of more than 130% from the previous year (See Fig. 1). Another APACS study showed that online banking fraud losses for the first semester of 2009 totalled £39 million — A 55% rise on the figure for the same period of 2008 [8].

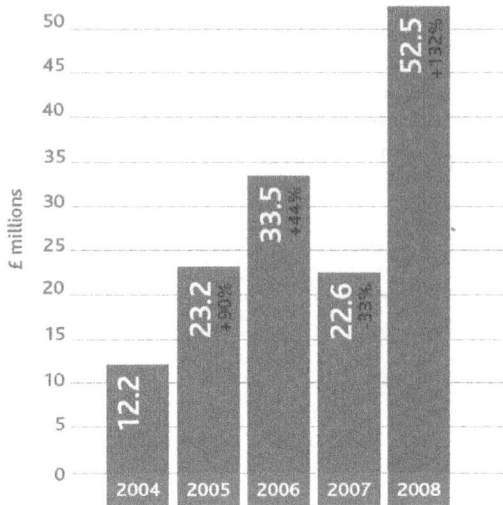

Fig. 1. Comparison of the annual losses to online banking fraud during 2004-2008. *Percentage values in grey* show the percentage change on previous year's total. [7]

Despite these alarming statistics, many banks encourage their customers to make use of online banking by claiming that their online banking websites are safe and secure. Studies by security experts such as Professor Atul Prakash and his students show otherwise. In 2006 they examined the websites of 214 financial institutions and found that 75% of the websites had at least one design flaw which cyber criminals could exploit to gain access to people's private information and bank accounts [9].

3   The Online Banking EFT Transaction Process

In an attempt to protect their clients from falling victim to online banking fraud, banks are integrating different security technologies into their online banking websites. A good example of this is two-factor authentication where banks require that a person needs to know the credentials (username and password pair) to gain access to an online bank account, as well as be able to provide a valid one-time password (OTP) when they wish to carry out certain transactions such as making a payment to a new beneficiary. An OTP is generally sent as an SMS message to the cell phone number of the account holder. The idea is that an account holder should be the only person who knows what the username and password to their account is, and that only they will have the cell phone (to which the OTP is sent) in their possession.

Let's examine the actions that are involved in a typical online banking payment transaction; also known as an electronic funds transfer (EFT) transaction. The following points describe how an account holder and registered online banking user would carry out an EFT transaction via the online banking website of a reputable bank in the Republic of South Africa. Please note however that the details of the steps in a typical EFT transaction may differ slightly from bank to bank.

1. The account holder (say John), fires up his browser, navigates to the bank's website, and selects the online banking hyperlink.
2. The website will ask John for his bank account number and the corresponding security PIN (Personal Identification Number).
3. If a valid bank account number and PIN pair is submitted then the website will ask the John for his online banking password.
4. When the correct online banking password is submitted, the bank will send an SMS message to John's cell phone number to notify him that someone has just logged into his bank account: "Confirmation of Internet Banking logon with acc no. ending 0101. Date: 2010-02-16. Time: 22:10:59. Helpline: 12345 12345. Int no.: +2711 123 4567"
5. John can now carry out several online banking transactions, e.g. view the transaction history of his account, pay beneficiaries, etc.
6. In order to make a payment to a new beneficiary, i.e. somebody he has never transacted before, the new beneficiary must first be added to the list of payable beneficiaries.
7. To register a new beneficiary John simply needs to click on the appropriate beneficiary-related option on the website menu. As a security measure the bank will then send a OTP (sometimes referred to as a Random Verification Number or RVN) to his cell phone number: "RVN — Enter a39c0c9f sent at 22:15:33 2010-02-16, to continue with your Internet banking session. Helpline: 12345 12345. Int no.: +2711 123 4567". The online banking website will ask John to enter an OTP into an input field on the screen. Only when the correct OTP has been submitted will John be allowed to proceed to the next step in the beneficiary registration process where the beneficiary's details, i.e. name, account number etc., can be supplied.
8. Once the new beneficiary has been registered, John will be able to make payments to the beneficiary. He simply needs to start an EFT transaction by selecting the appropriate option on the website menu, then provide details for the transaction (i.e. beneficiary to pay, amount to transfer, payment date etc.), and finally confirm that the transaction details that was submitted are correct.

At first glance the transaction process looks fairly secure. How could a cyber criminal possibly circumvent all the security barriers described above? As the following sections will explain, this is all in a day's work for an experienced fraudster.

4   Attack Vectors

From the discussion above we can identify 4 items of information that a fraudster would need to access an online banking user's account and carry out EFT transactions:

- The account number.
- The security PIN for the account number.
- The account holder's online banking password.
- The OTP which is sent to the account holder's cell phone number.

The first 3 items of information are fairly easy for a seasoned cyber fraudster to acquire — With local and remote attack vectors such as social engineering, phishing and spyware, online banking users can be duped to unwittingly divulge the secret information. Obtaining the final item of information (the OTP) however, is a whole new ball game. An OTP is randomly generated and is only valid for a single online banking session. The moment the online banking user logs out or closes their browser the OTP expires. Furthermore, the OTP is only sent to a single destination, namely the account holder's cell phone number. A fraudster will therefore have to find a way to intercept the OTP if he has any hope of successfully defrauding his target. But how would he accomplish this seemingly impossible task without the account holder's knowledge or consent? Enter "SIM swap fraud".

5   SIM Swap Fraud

"A Western Cape man was defrauded of more than R100 000 shortly before Christmas after he fell victim to a Joburg syndicate that illegally swopped his cellphone SIM card so that they could access his bank account. Barry Greyvenstein, from Grootbrak River in George, said he realised that something was amiss when he received a call from Absa Bank on the evening of Friday December 7 informing him of irregular activity on his bank account. But when he checked his phone, there was no signal - and he had become the latest victim of a new type of fraud, which earlier resulted in fraudsters plundering R90 000 from the bank account of a Cape Town NGO. This week Greyvenstein, who mainly uses internet and cellphone banking for transactions, said he was blown out of my boots' when he realised that MTN had performed a SIM card swop on his number without his knowledge or consent." [10]

5.1   What Is SIM Swap Fraud?

SIM swap fraud is a cunning scam where fraudsters hijack a targeted online banking user's cell phone number in order to obtain the OTPs and security messages that the account holder's bank would send to the cell phone number during online banking transactions [11]. SIM swap fraud has in a very short period of time become a source of major concern for mobile network operators, banks and cell phone users alike. According to a report released by the e-commerce unit of the South African Police Service (SAPS) more than R80 million has been lost to SIM swap fraud in South Africa since February 2007 [12].

The scam works as follows:

- **Step 1: Gathering information**
  A fraudster may select a specific person as a target, or choose to target a random group of people with the hope that he will successfully defraud a few of them. Whatever the case, the first step entails collecting personal and confidential information about his target(s). Information of interest to the fraudster includes ID numbers, contact details, residential and postal addresses, banking details such as account numbers, credit card numbers, and online banking credentials (username and password). This is rather sensitive information which people will not willingly hand over to a just anybody, let alone a complete stranger who happens to be a fraudster. The fraudster will therefore have to make use of social engineering and phishing scams to trick his target(s) into disclosing their precious personal information.

- **Step 2: Requesting a SIM swap**
  Mobile network operators like Vodacom, MTN and Cell C know how inconvenient it is for their clients to lose their cell phone numbers when their SIM card is lost or damaged, or their cell phone is stolen. For this reason they offer a SIM swap service, which allows their clients to request a SIM card swap on their cell phone number in order to replace a lost, stolen or damaged SIM card while keeping their original cell phone number. When a SIM swap is performed, the old (lost, stolen or damaged) SIM card is disabled and the client's cell phone number is linked to a new (replacement) SIM card. All that the mobile network operators ask in return is a small fee, and that the person requesting the SIM swap can prove that they are the owner of the cell phone number by providing an identity document and correctly answering a set of personal-information-related security questions.

  Fraudsters exploit this SIM swap service that mobile network operators provide, to hijack their targeted victims' cell phone numbers. By masquerading as a targeted victim, a fraudster will ask a mobile network operator to transfer the victim's cell phone number to a SIM card in his possession. The information that was collected in step 1 is used to help convince the mobile network operator that the SIM swap request is "legitimate", i.e. that the fraudster is the lawful owner of the cell phone number.

- **Step 3: Looting the victim's bank account**
  When the SIM swap is complete, it is a race against the clock for the fraudster to loot his target's bank account. He must act quickly, before the victim notices that their cell phone is no longer connected to a cell phone network, and therefore unable to receive phone calls and SMS notifications. The fraudster will use the credentials that were acquired in step 1 to log into the online banking website of the bank where the victim has an account. Upon successful login, the bank will send an SMS notification to the account holder's cell phone number to inform him/her about it, but because the fraudster has hijacked his/her cell phone number, the notification will be delivered to the fraudster's cell phone instead. The fraudster will then proceed to add one

or more beneficiary accounts to the victim's bank account. To authorize the addition of any new beneficiaries, the bank will request an OTP which will be delivered via an SMS notification to the hi-jacked cell phone number. At this point the fraudster will be smiling as after entering the OTP, he can proceed to transfer funds into each of the beneficiary accounts, and finally, disappear into the vastness of cyberspace.

6 Problem Analysis

From the discussion above it is clear that there are 4 parties involved in a case of SIM swap fraud, namely:

- The fraudster
- The victim
- The victim's mobile network operator
- The victim's bank

While the fraudster escapes with the money, the latter 3 parties find themselves in a situation where they point a finger of blame at one another — They do not want to accept responsibility for their contribution to the fraudster's success in the scam with which they were fooled.

Online banking users — and Internet users in general — are expected to take best efforts to protect their personal information. Banks regularly warn their clients that they will never send them an SMS or e-mail which will ask them to click on a hyperlink that will take them to a website where they must update or confirm their online banking details. In spite of these warnings, there are people who still fall into a fraudster's phishing trap because the phishing e-mail/SMS is so cleverly designed and phrased that they believe it is an authentic e-mail/SMS from their bank. The victim is therefore guilty of compromising his/her personal information and online banking credentials when he/she fell prey to the fraudster's phishing scam(s).

Is it fair though to put full blame on the victim? Surely the bank cannot be allowed to get away scot free because it displays security notifications to its clients from time to time. The bank is after all responsible for protecting its clients' money from criminals. But the bank's online banking system failed to recognize that it was not the true account holder that was logged in (Which begs questions about the effectiveness of the bank's online banking authentication system), and that there was abnormal behaviour (i.e. transactions) on the account. Consequently the fraudulent transactions were approved and the victim is left with a big void in his/her bank account.

Finally, the mobile network operator is guilty of carrying out a fraudulent SIM swap after failing to ascertain that the person who requested the SIM swap was not who they claimed to be. A forged or stolen identity document and the correct answers to a pre-known set of security questions was all the fraudster needed to deceive the mobile network operator and hijack the victim's cell phone number. Paper-based identification documents are no longer good enough — The speed of technology has far outpaced the security of countries' identity documents [13].

7 Proposed Solution

To effectively combat SIM swap fraud, focus will have to shift from protecting on-line banking users' personal information, to the second and most critical link in the SIM swap fraud scam: preventing unauthorized SIM swaps. As discussed earlier, a successful SIM swap is key to a fraudster's success as it allows the fraudster to intercept the OTP messages from a victim's bank. If a good security system is put in place here, the advantages will be twofold: cell phone subscribers will enjoy the peace of mind that their cell phone numbers are safe from being hijacked, and SIM swap fraudsters will be at bigger risk of getting caught and brought to book.

Enter "BIO-Swap" — Our proposed solution to address the issue of SIM swap fraud. Short for "Biometric-Swap", BIO-Swap is a biometrics- and Web-based proof-of-concept system which is designed to serve as a type of certification authority for people's identities. The idea is that BIO-Swap will vouch for registered cell phone subscriber's identity when they request a SIM swap from their mobile network operator. In layman's terms the BIO-Swap system will act as a trusted third party when SIM swaps are conducted: the mobile network operator will trust BIO-Swap to accurately verify whether a person is the legitimate owner of a given cell phone number, and the cell phone subscribers will trust BIO-Swap to protect their cell phone numbers from fraudsters.

8 The BIO-Swap System in a Nutshell

As mentioned earlier, BIO-Swap is a biometrics-based system. It uses the power of biometrics technologies to capture a person's biometrics, and to extract a set of biometric templates thereof. The acquired biometric templates are then linked to the person's cell phone number as a hypothetical biometric lock which purpose is to protect the cell phone number against unauthorized SIM swaps.

Why did we opt for biometrics to replace traditional identification documents and security questions? Because a biometric of a person is a characteristic or trait of that person which distinguishes him/her from the other people on earth; the probability of 2 people sharing the same biometric data is virtually zero. Furthermore, biometric properties are extremely difficult to duplicate or share as they are intrinsic properties of the owner. Individuals can therefore be [uniquely] identified by their biometrics. [14]

A cell phone subscriber will need to register for the "BIO-Swap SIM Swap Service" if they wish to enjoy the protection that the BIO-Swap system has to offer. The registration process is quick and painless, and is carried out under the watchful eyes of a BIO-Swap supervisor who will assist and guide the cell phone subscriber through the process:

1. Say Mandy wishes to register for the BIO-Swap SIM Swap Service. The supervisor will first log into the BIO-Swap user interface and then proceed to ask her for a few items of personal information such as her name, surname, and a certified copy of her identity document. In addition, Mandy will have to provide cell phone-related information such as her cell phone number, contract number (if applicable) and her SIM card number.

2. Once the required information has been collected, the BIO-Swap system will connect to — and communicate with — the computer systems of Mandy's mobile network operator to verify whether the information that she provided (think contract number, SIM card number and personal details) is correct. This also serves as a notification to the mobile network operator that Mandy is attempting to register for the BIO-Swap SIM Swap Service.

3. If all the information that Mandy provided is correct and the mobile network operator has no objections to the BIO-Swap registration attempt, it will respond by sending an OTP via SMS to the cell phone number. This one-time password serves as a challenge to verify whether Mandy is the truly the owner of the cell phone number she is attempting to register for the BIO-Swap SIM Swap Service, or if she is a fraudster who has collected the information by means of a phishing scam. In other words, Mandy needs to prove that she has the cell phone number (i.e. SIM card to which the cell phone number is linked) in her "possession".

4. Say Mandy is the lawful owner and she receives the OTP from the mobile network operator. She will need to provide the OTP to the supervisor, who will then enter it in an "authorization code" input field on the user interface and submit it to the mobile network operator.

5. If the submitted authorization code is correct, the mobile network operator will approve the registration request and give the BIO-Swap system a green light to proceed to the next step of the registration process, namely biometric enrollment. Otherwise, after 3 failed attempts to provide the correct OTP, the registration process will be terminated.

6. Say the correct OTP was submitted. The BIO-Swap system will now require Mandy's biometrics in order to complete the registration process. When the BIO-Swap biometric enrollment Java applet has finished loading into the browser, the supervisor will help Mandy to scan her biometrics with a bio-metrics scanner/reader. The prototype system that we have developed uses a fingerprint scanner to capture images of a random subset of a subscriber's fingerprints (Fig. 2).

7. When the required number of fingerprint images have been captured, the supervisor must authorize the captured biometrics with his password and a scan of one of his own fingerprints (which he enrolled when his supervisor user account was created). This serves as an assertion to the BIO-Swap system that he oversaw the capturing of Mandy's biometrics and that, to the best of his knowledge, there was no foul play involved.

8. The applet will then extract biometric templates from all of the acquired fingerprint images and encrypt the resulting byte arrays with the AES cipher [15], using the MD5 hash [16] of the password that the supervisor entered as a 128-bit encryption key. At this point there is no relationship between the original fingerprint images and the encrypted fingerprint template byte arrays, so the subscriber's biometric data is safe for transmission over the public Internet (I.e. there is no way to recover the fingerprint images from the encrypted biometric templates). In the event of a man-in-the-middle attack, the biometric data will be worthless to the interceptor.

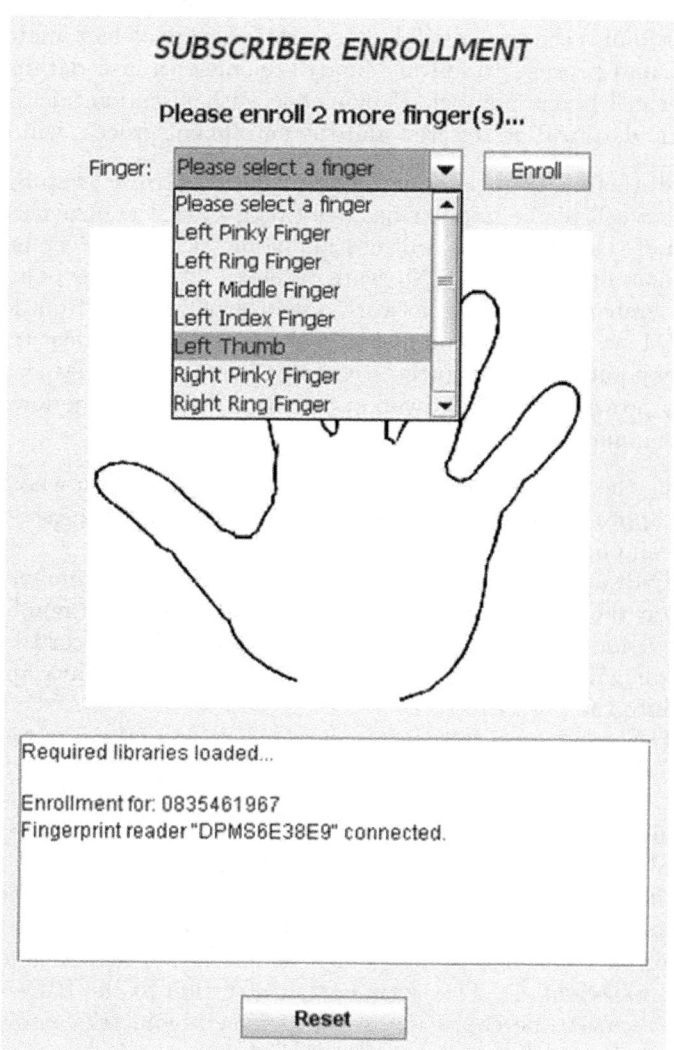

Fig. 2. The supervisor must select a finger to enroll

9. The applet will now send the encrypted biometric data along with some meta-data to a remote BIO-Swap server. When the BIO-Swap server receives the data, it will retrieve the supervisor's password hash from a database, use it decrypt the biometric data, and then attempt to match the fingerprint template (that was captured in step 7 above) to one of the templates it has on record for the supervisor. If biometric authentication is successful (i.e. a match is found) the server will be assured that this is a legitimate enrollment request, and proceed to enroll Mandy's biometrics in a database and link it to her cell phone number. If biometric authentication fails however, the biometric data will be rejected and the enrollment process will fail.

That is all there is to registering for the BIO-Swap SIM Swap Service. From the moment a cell phone number has been registered, the owner can enjoy the peace of mind that BIO-Swap will use his biometrics to protect his cell phone number against unauthorized SIM swaps. When a SIM swap is requested on the cell phone number, the mobile network operator will refrain from following the standard SIM swap process, and instead entrust the task of identity verification and SIM swap authorization to the BIO-Swap system. The SIM swap requestor will have to prove to the BIO-Swap system that he/she is the lawful owner of the cell phone number:

1. Following the use case described earlier, let's say a person who claims to be Mandy approaches her mobile network operator and requests a SIM swap on her cell phone number.
2. The BIO-Swap supervisor will search for the cell phone number or Mandy's details on the BIO-Swap system to see if the cell phone number has been registered for the BIO-Swap SIM Swap Service. If a record is found, the supervisor will select it and the BIO-Swap SIM swap Java applet will be loaded into the browser.
3. The BIO-Swap system will select a random subset of the fingers that Mandy had enrolled when she registered for the BIO-Swap SIM Swap Service. The person, who is requesting the SIM swap on her number, must now scan these fingers with the provided fingerprint scanner, under the supervision and guidance of the BIO-Swap supervisor (Fig. 3).
4. When the person has scanned all the fingers the applet asked for, the supervisor must authorize the captured biometrics with his password and a scan of one of his own fingerprints (which he enrolled when his supervisor user account was created). This serves as an assertion to the BIO-Swap system that he oversaw the capturing of the person's biometrics and that, to the best of his knowledge, there was no foul play involved.
5. As described earlier in step 8 of the registration process, the applet will create biometric templates from the acquired fingerprint images where after it will encrypt them for secure transmission over the public Internet.
6. Finally, the applet will now send the encrypted biometric data along with some meta-data to a remote BIO-Swap server. When the BIO-Swap server receives the data, it will perform biometric authentication on the supervisor's fingerprint, exactly as described earlier in step 9 of the registration process.

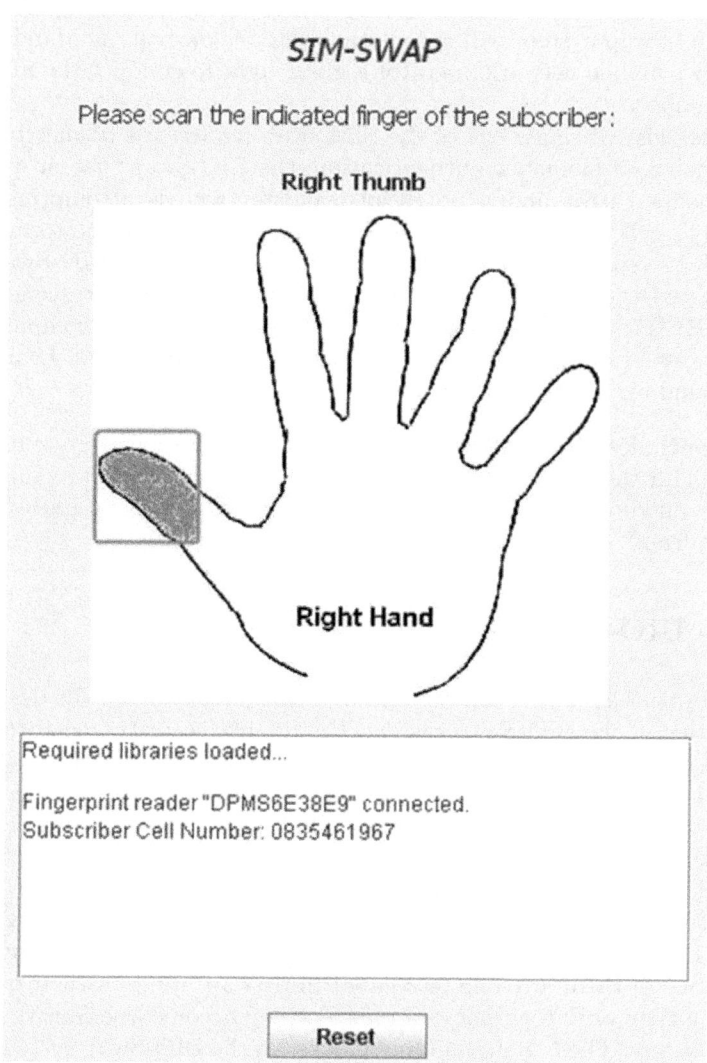

Fig. 3. The supervisor must scan the indicated fingers of the person who is requesting the SIM swap

7. If biometric authentication of the supervisor's fingerprint is successful, the server will proceed to perform biometric authentication on the SIM swap requestor's fingerprints — It will retrieve Mandy's fingerprint templates from a database and compare it to the SIM swap requestor's fingerprint templates. If a match is found for each of the SIM swap requestor's fingerprint templates, the BIO-Swap system will accept that the person truly is Mandy, and give Mandy's mobile network operator a green light to conduct the SIM swap on her number.

   Otherwise, if single one of the SIM swap requestor's biometric templates does not pass biometric authentication, the BIO-Swap system will consider him/her a threat and a potential fraudster who is attempting to hijack Mandy's cell phone number. The SIM swap process will be terminated and the SIM swap will consequently be denied. In addition, if BIO-Swap has been integrated with the computer systems of law enforcement agencies such as the FBI [17], the person's biometric templates could be compared against databases of biometrics of known criminals and fraudsters. If any matches are found authorities can be alerted.

This concludes the high-level overview of the BIO-Swap system. The next section will lift the hood of the BIO-Swap system to reveal what its main building blocks are, and how they work together to achieve its ambitious goal of combating SIM swap fraud.

9   The BIO-Swap System Architecture

The authors put a lot of effort, thought and consideration into the design and implementation of the BIO-Swap system. Factors like security [most importantly], efficiency, usability, robustness, availability and maintainability were high on the agenda. We visualized BIO-Swap from each of its end-users' perspective to identify what they would expect from it, and how it could win their trust. Furthermore, some of the latest (at the time of this writing) software frameworks were used to develop the system.

The result of the above was a prototype of a multi-part Web-based system which consists of 5 main components, each of which can function independently from the others. None of them are able to combat SIM swap fraud on their own though. Only when they work together as a unit, do they become a serious threat to SIM swap fraudsters. The 5 main building blocks[1] of the BIO-Swap system are:

- **The BIO-Swap Web application:** A Microsoft ASP.NET 3.5 Web application. It is the main graphical user interface (front-end) which authorized BIO-Swap users can use to interact with the BIO-Swap system and to access the functionality it has to offer. Depicted as "A" in Fig. 4.

---

[1] BIO-Swap is a fairly large and complex system. A detailed and technical discussion of its 5 main components was not possible in this paper due to the constraint on the allowed number of pages. Please contact the authors for more information.

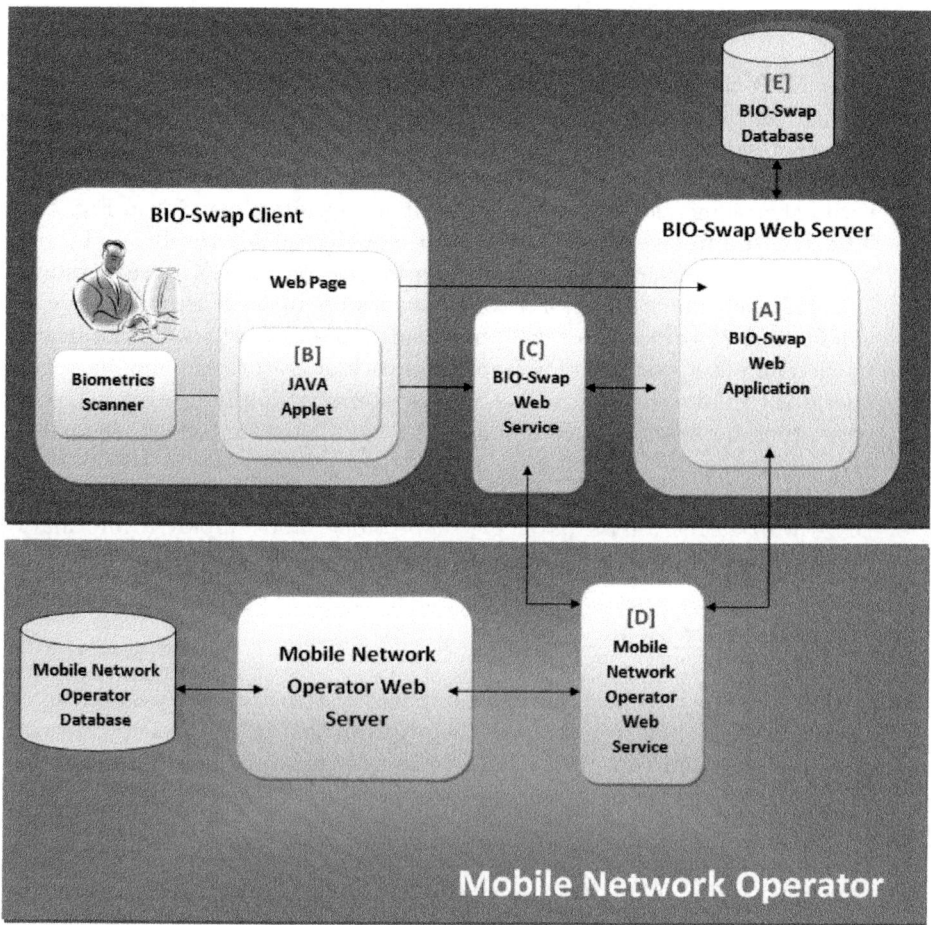

Fig. 4. The BIO-Swap architecture.The *arrows* indicate in what directions data flows between the components.

- **The BIO-Swap Java applet:** A Java applet that gets embedded in some of the Web pages of the BIO-Swap Web application. It runs inside a client's browser, and its main functions are client-side biometrics acquisition and biometric templates extraction. In addition, the applet consumes the BIO-Swap Web service (discussed below) and calls upon its services when biometric data needs to be sent to the BIO-Swap Web server for processing (i.e. biometric verification during SIM swap requests) or storage in the database (i.e. biometric enrollment during subscriber registration). Depicted as "B" in Fig. 4.
- **The BIO-Swap Web service:** A Microsoft .NET Framework 3.5 Web service that serves as an interface between the BIO-Swap Web server and the client-side Java applet (discussed above). It exposes services to enroll the biometric templates of new subscribers into the BIO-Swap database, and to verify the biometrics of existing subscribers. Depicted as "C" in Fig. 4.
- **The mobile network operator Web service:** A Microsoft .NET Framework 3.5 Web service which provides a means for the BIO-Swap system to exchange information with a mobile network operator. It exposes services to validate the information of cell phone subscribers, and to send notifications to the computer systems of a mobile network operator. Any mobile network operator can integrate with BIO-Swap system by simply hosting a Web service which implements a common BIO-Swap-defined interface. Depicted as "D" in Fig. 4.
- **The BIO-Swap database:** A Microsoft SQL Server 2005 database in which BIO-Swap stores all its data, e.g. subscriber- and employee information, biometric templates, etc. Depicted as "E" in Fig. 4.

10   Pros and Cons

The positive and negative aspects of a computer system are often weighed against each other to assess how well it solves the problem that it was designed for. Let's compare the pros and cons of the BIO-Swap system to see how it fares.

10.1   Pros

On the positive side, BIO-Swap offers the following advantages to its users:

- It is safe and anonymous in the sense that no biometric images are ever stored on a hard-drive or transmitted over the public Internet — Only biometric templates are stored and transmitted.
- It is completely flexible with regards to biometric technologies because it is biometric template driven. Other forms of biometrics such as facial recognition, hand-, iris-, and retina biometrics are supported and can be easily integrated.
- It is completely Web-based. This allows for the effortless and centralized distribution of Web pages and necessary client-side software (e.g. the BIO-Swap Java applet) to remote computers.

- Biometric enrollment and SIM card swaps are carried out in a protected and controlled environment, under the watchful eyes of a trained supervisor. Successful biometric identification and verification within this controlled environment is a prerequisite of a SIM swap on a cell phone number that has been registered for the BIO-Swap SIM Swap Service.
- It relieves mobile network operators of the difficult task of verifying whether a person is the legitimate owner of a cell phone number when a SIM swap is requested, and will accept accountability if fraudster manages to carry out an unauthorized SIM swap on a registered cell phone.

10.2 Cons

There is unfortunately no such thing as a 100% perfect and secure computer or software system [18]. The BIO-Swap system is no exception. Like every other man-made creation, it is not flawless:

- It is dependent on stable Internet connections because it is a Web-based system. Any network problems will therefore render the system non-functional, which could be a problem when an emergency SIM swap needs to be carried out.
- By using biometric technology, BIO-Swap inevitably inherits all the cons associated with it. This includes:
  - False acceptance rate (FAR). I.e. on rare occasions, the system may make a positive match between a scan of one person's fingerprint and another person's biometric template in its database.
  - False rejection rates (FRR). I.e. the system may occasionally fail to find a positive match between a scan of a person's fingerprint and their biometric template in its database.
  - On rare occasions the system may fail to enrol a person's biometrics because it is of very poor quality or badly damaged. Think of construction workers who have little or no fingerprints because the hand labour they do has worn it away.

From the points discussed above it is clear that the positive aspects of the BIO-Swap system far outweigh its negative aspects. The issues that are inherited from biometric technology will become less of a problem as the technology (i.e. biometric scanners, algorithms, etc.) improves over time. As for the dependency on stable Internet connections: Internet downtime is out of the control of the BIO-Swap system, but when it does occur it is usually resolved quickly because it is equally critical to the operation of many other organisations and computer systems. It is therefore fair to say that BIO-Swap has the potential to effectively reduce — maybe even eliminate — SIM swap fraud.

11   Summary and Conclusion

This paper started off with a discussion of online banking: the Internet-based service which evermore people across the world are adopting because it allows them to quickly and conveniently do their bank transactions from just about anywhere in the world. As we saw from a few statistics however, this ease-of-use and convenience comes at a price. Online banking users are vulnerable to sophisticated online banking fraud scams that are master-minded and executed by cyber fraudsters. We saw that total annual losses to online banking fraud in the UK alone exceeded £50 million in 2008.

To determine what attack vectors a cyber fraudster could possibly use to defraud a victim, the online banking EFT transaction process was examined for vulnerabilities. We discovered that the mechanisms that banks have put in place to secure online banking transactions (at the time of this writing), can be circumvented via SIM swap fraud — A cunning scam where fraudsters hijack targeted victims' cell phone numbers in order to intercept the OTP messages that are required to authorize fund transfers to new beneficiaries. An analysis of what SIM swap fraud entails revealed that fraudsters are able to carry out illegal SIM swaps because mobile network operators rely on a set of security questions and identification documents to verify a subscriber's identity.

The paper then introduced a biometrics-based security system called "BIO-Swap" which the authors have developed to combat SIM swap fraud. We saw that BIO-Swap is designed to act as a certification authority for cell phone subscribers' identities, with the goal to prevent unauthorized SIM swaps. A brief run-through of the BIO-Swap registration- and SIM swap process was given to explain how BIO-Swap works, followed by a high-level overview of its main components. Finally, the positive and negative aspects of the BIO-Swap system were discussed.

To conclude: Nothing is impossible for determined cyber criminals, but with the right tools and well-designed security systems we can make their lives extremely difficult. So much so that whatever they have to gain from their nefarious deeds will not be worth the effort and risk of getting caught.

References

1. Fiserv Survey Shows Online Banking Growing, Now Used by Four of Five Online Households,
   http://investors.fiserv.com/releasedetail.cfm?ReleaseID=396336
2. Half of UK net users bank online,
   http://www.finextra.com/news/fullstory.aspx?newsitemid=20938
3. South Africa Internet Usage and Marketing Report,
   http://www.internetworldstats.com/af/za.htm
4. Social networking in South Africa,
   http://mybroadband.co.za/news/Internet/11238.html
5. Jesse James Biography,
   http://www.biographybase.com/biography/James_Jesse.html
6. History of Butch Cassidy, LeRoy Parker,
   http://www.utah.com/oldwest/butch_cassidy.htm
7. Fraud the Facts 2009: The Definitive Overview of Payment Industry Fraud and Measures to Prevent it. APACS, London (2009)

8. Financial Fraud Action UK announces latest fraud figures,
http://www.banksafeonline.org.uk/documents/
2009H1FraudPressRelease.pdf
9. Security flaws in online banking sites found to be widespread,
http://www.ns.umich.edu/htdocs/releases/story.php?id=6652
10. Beware SIM card swop scam,
http://www.security.co.za/fullStory.asp?NewsId=5907
11. Protect yourself from fraud,
http://www.standardbank.mu/portal/site/mauritius/
menuitem.cb169d81ccc6cb0e7b6965103367804c/
?vgnextoid=c3876ddd47d51210VgnVCM10000050ddb60aRCRD
12. Police probe SIM swap fraud,
http://www.mydigitallife.co.za/index.php?option=com_content&
task=view&id=1036940&Itemid=38
13. Fake IDs, Fake Passports Easy To Make or Buy,
http://realtysecurity.com/blog/2009/03/16/
fake-ids-fake-passports-easy-to-make-or-buy/
14. Biometric systems offer many important benefits,
http://www.biometricnewsportal.com/biometrics_benefits.asp
15. The AES-CBC Cipher Algorithm and Its Use with IPsec,
http://w3.antd.nist.gov/iip_pubs/rfc3602.txt
16. The MD5 Message-Digest Algorithm, http://www.ietf.org/rfc/rfc1321.txt
17. Federal Bureau of Investigation, http://www.fbi.gov/
18. Kaspersky, E.: no such thing as 100% secure software,
http://www.pcadvisor.co.uk/blogs/index.cfm?entryid=104702&blogid=4

Are BGP Routers Open to Attack?
An Experiment

Ludovico Cavedon, Christopher Kruegel, and Giovanni Vigna

University of California,
Santa Barbara
{cavedon,chris,vigna}@cs.ucsb.edu

Abstract. The BGP protocol is at the core of the routing infrastructure of the Internet. Across years, BGP has proved to be very stable for its purpose. However, there have been some catastrophic incidents in the past, due to relatively simple router misconfigurations. In addition, unused network addresses are being silently stolen for spamming purposes. A relevant corpus of literature investigated threats in which a trusted BGP router injects malicious or wrong routes and some security improvement to the BGP protocol have also being proposed to make these attacks more difficult to perform. In this work, we perform a large-scale study to explore the validity of the hypothesis that it is possible to mount attacks against the BGP infrastructure without already having the control of a "trusted" BGP router. Even though we found no real immediate threat, we observed a large number of BGP routers that are available to engage in BGP communication, exposing themselves to potential Denial-of-Service attacks.

1 Introduction

The *Border Gateway Protocol* (BGP, [19]) is the routing protocol at the core of the Internet. BGP is employed to perform routing decision between *Autonomous Systems* (ASes), which are separate administrative domains. Directly connected Autonomous Systems establish *peering* relationships. A peering relationship implies full trust: one router will accept and use for routing any network prefix advertised by its peer routers (unless administratively forbidden).

This full trust between peers is one of the weaknesses of the protocol. In fact, unconditional acceptance and propagation of routing information coming from other peers might render the whole Internet routing stability vulnerable to malicious, compromised, or just misconfigured BGP routers. According to Mahajan et al. [13], BGP misconfiguration is quite common (up to 1% of the global BGP table entries), although only 4% of these misconfigured announcements result in disrupted connectivity. Nevertheless, wrongly advertised prefixes sometimes gave place to well-known catastrophic routing incidents. For example, Pakistan Telecom disrupted worldwide routing to YouTube's web site in an attempt to prevent access to that web site in Pakistan [20]. Another example is the AS7007 incident [2], where the wrong propagation of routes caused an Internet-wide

J. Camenisch, V. Kisimov, and M. Dubovitskaya (Eds.): iNetSec 2010, LNCS 6555, pp. 88–103, 2011.

blackout. In addition, an AS might generate malicious BGP prefix advertisements in order to hijack some IP addresses and use them as not-yet-blacklisted sources of spam, as reported by Ramachandran and Feamster [18] and analyzed by McArthur and Guirguis [14].

Previous literature has been investigating possible areas of attack against the BGP protocol (see, for example, [15] and [3]). Pilosov and Kapela [17] also gave a live demonstration about the feasibility of an unnoticed *Man-In-The-Middle* (MITM) attack performed by means of BGP prefix hijacking.

Much work has focused on finding solutions to the above-mentioned attacks. For example *S-BGP* ([11]), *soBGP* ([10]), and *IRV* ([8]) are BGP extensions aimed at authenticating prefix origins and updates. However, most of the proposed solutions imply a heavier load on the CPU and memory of routers in order to perform cryptographic operations. Moreover, changes to the protocol need to be backward compatible, as it not possible to replace all the routers software at once. For these reasons, adoption of the defense techniques proposed so far has been extremely slow.

Almost all of the successful attacks against BGP considered in the literature require the attacker to have control of a BGP router. This condition, however, is not easy to achieve and maintain, while having a malicious behavior. In fact, the attacker has to demonstrate the need to become an AS (i.e., at least being a multi-homed network), or must be supported by the complicity of an ISP (which is willing to accept the risk of damaging its own reputation), or needs to compromise an existing BGP speaker (but router exploits are rather rare), or requires to hijack an existing TCP connection between two BGP routers (which is very hard to carry out).

However, there is an unanswered question: Is it possible to mount attacks in order to disrupt inter-domain routing without already having the control of a "legitimate" BGP router? In this work, we perform a large-scale study to explore the validity of this hypothesis.

First, we tried to identify how many BGP speakers were reachable on the Internet, by performing a SYN scan for TCP port 179 over a very large part of the Internet address space (about 73%). The scan was performed between December 2008 and January 2009. 2.2 million hosts (0.8‰) answered to our SYN packets, identifying an equal number of processes listening on that TCP port. Clearly, there is no implication that those hosts were BGP speaker. Restricted to this subset of IP addresses, two additional scans (performed in February 2009 and February 2010) went further in the connection negotiation, aiming at detecting whether the counterpart was willing to continue after a 3-way handshake and even establish a BGP session.

In the rest of this paper, we present our findings and we describe the issues we encountered with such a large-scale scanning experiment.

2   BGP Basics

The *Border Gateway Protocol* (BGP) is specified in RFC 4271 [19]. BGP is a path-vector routing protocol used across *Autonomous Systems* (ASes) on the

Internet. BGP routers exchange information about reachable destinations (under the form of CIDR IP prefixes) with an associated path to traverse in order to reach them. The information collected by the BGP process is employed to perform routing decisions between ASes. In most setups, the BGP routers of an AS exchange prefix vectors only with the BGP routers of neighbouring ASes (i.e., directly connected ASes). In such cases, the two BGP routers are said to be in a *peering* relationship. Peering relationships between routers of two different AS are also called *eBGP* (*external* BGP) peerings, in order to differentiate them from connections between BGP routers within the same AS, called *iBGP* (*internal* BGP) peerings. A peering relationship implies full trust: one router will accept and use for routing any network prefix advertised by the other router (unless administratively forbidden).

Peering relationships are carried out using a TCP connection on port 179. Upon establishment of such connection, the two BGP routers exchange an *OPEN* message, which contains identification parameters like the AS number and the BGP identifier (usually the IP address assigned to one of the network interfaces) of the sender.

In case a BGP speaker accepts the peering request (on the basis of the remote IP address of the TCP connections and the information contained in the *OPEN* message), it sends to the remote party a series of *UPDATE* messages, containing the description of the prefixes that the router is configured to advertise. At the same time, the router listens for route updates form the other peer, and stores them in its *Routing Information Base* (RIB) according to the configured policy.

The lifetime of the peering relationship is bound by the TCP connection: if such connection is closed or lost, all the routing information learned via that connection is removed from the RIB. The connection is checked for liveliness with periodic *KEEPALIVE* messages.

Any error condition, for example during the *OPEN* negotiation or in *UPDATE* messages, is notified by sending a *NOTIFICATION* message, and the TCP connection is subsequently closed. Instability of a BGP connection might cause a potentially big set of routes to be cyclically added and removed from the RIB. This would have the effect of increasing the load on the router and on the network (as route *UPDATEs* would be consequently propagated). In order to overcome this problem, *Route Flap Dampening* (RFD, [22]) has been defined, which temporarily suppresses unstable routes. As it will be discussed in the next section, an attacker can exploit this functionality in order to disrupt connectivity.

In some cases, it may be required to establish an eBGP peering between two routers that are not directly connected, creating so called *multi-hop* BGP sessions. This may be the case, for example, when the two peers communicate via a device that does not support BGP or via links for which static routes have been set.

The main authentication mechanism for BGP is via the remote IP address, the router identifier and the AS number. Clearly this information is not enough to defend against spoofing attacks. BGP also provides an *Authentication* option in the *OPEN* message, which would allow for remote peer authentication, but would not defend against TCP-level attacks. Moreover, this option has never found widespread usage. In fact, in the latest BGP specifications the *Authentication* options has been deprecated in favor of TCP MD5 signatures [9]. When two peers are communicating using TCP MD5 signatures, every TCP packet they exchange must carry a specific TCP option containing an MD5 hash of the packet concatenated with a shared secret. The value of the shared secret must be manually specified in the configuration of both routers. In this way, an attacker who does not know the secret cannot establish a spoofed peering session or hijack/reset an existing TCP connection, as non-signed packets are discarded.

3   Attacks against BGP

The biggest weakness of BGP is the trust that a router has toward its peers (see Fig. 1). Any route that is advertised by a neighboring BGP speaker is merged in the routing database and is propagated to all the other BGP peers. Configuration mechanisms for filtering incoming and outgoing prefix advertisements are available, but clearly having a system administrator manually specify acceptable prefixes mostly defeats the purpose of having a routing protocol like BGP. Moreover, the number of prefixes a BGP router is handling might be extremely large: the current the size of the *Forwarding Information Base* (FIB) of a BGP router is above 300,000 entries[1].

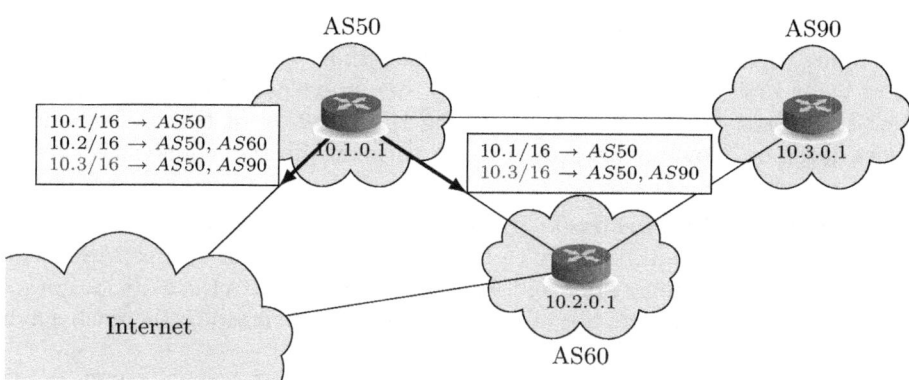

Fig. 1. Example of BGP prefix advertisement propagation. Two prefix advertisement messages sent from the router of Autonomous System AS50 are represented in the picture. AS50 is claiming to be able to reach the network 10.3.0.0/16 by being directly connected to AS90. AS60 and the other ASes on the Internet need to trust this piece of information from AS50.

3.1   Attacks from a Trusted Peer

Exploiting a trust relationship, a malicious or misconfigured router can claim ownership or reachability of network prefixes it is not entitled to.

---

[1] `http://bgp.potaroo.net/as1221/bgp-active.html`, as of February 2010.

Such advertised prefixes will then be spread potentially to all BGP routers and used to build routing tables, having the effect of directing IP packets towards the misbehaving router rather than the legitimate AS.

Hijacking traffic can be performed for multiple reasons. For example, an entity or malicious AS may want to make a subnet unreachable from other Internet hosts, or "blackhole" it. In this case, such AS can start advertising the prefix for the target network and then discard all the traffic it receives that is intended for that network. This is what happened in February 2008 in Pakistan [20], where the government decided to block access to the YouTube website, by having the main Pakistani telecom provider hijack YouTube network prefixes. However, a BGP misconfiguration caused the hijacked prefixes announcements to be propagated outside Pakistan boundaries, therefore blackholing YouTube's website across the whole Internet.

Attacks to the BGP infrastructure can be more sophisticated than blackholing. The ability of diverting traffic toward one's own AS allows for mounting *Man-In-The-Middle* (MITM) attacks, where traffic is sniffed or even modified. Pilosov and Kapela [17], in fact, gave a live demonstration about the feasibility of attracting most of the traffic destined to a target AS and than route it back to the intended AS, without users noticing it, and hiding the presence of the attacker from the output generated by routing analysis tools, such as Traceroute.

Probably, the most common reason for willingly stealing a network IP prefix is for spam purposes. Ramachandran and Feamster [18] observed that a percentage of spam emails (growing up to 10% from time to time) are sent from IP addresses belonging to short-lived prefixes (i.e., prefixes that are withdrawn within a day after being firstly advertised). In this way, spammers can obtain non-blacklisted new IP address to use; moreover keeping the duration of the prefix hijacking short does not allow the affected ASes to perform effective countermeasures.

Effectively stealing prefixes. If two ASes advertise the same prefix, the ASes across the Internet will be likely to be partitioned into two groups: ASes which are closer to the legitimate one and ASes which are closer to the malicious or misconfigured one. Therefore, not all the traffic will be diverted to the misbehaving router.

In order to increase the effectiveness of an IP prefix hijacking attack, a commonly used technique is to advertise subnets with a longer netmask than the original netmask of the prefix to be hijacked. As BGP will build forwarding rules using the longest-prefix matching criteria, the newly advertised rules will override the original one.

On the other side, the longer prefix matching functionality can also be exploited by advertising shorter prefixes: in this way the advertiser will steal unused subnets without disrupting the routing of actively-used prefixes.

3.2   Attack Which Do Not Need Control of a Trusted BGP Router

If an attacker does not control a trusted peer, some other attacks are possible at the TCP level. First of all, a malicious node able to sniff traffic between two BGP peers could be able to hijack the TCP connection and inject its own *UPDATE* messages. Achieving favorable conditions for this attack, however, is difficult to achieve, given that usually eBGP routers are connected via dedicated point-to-point links.

BGP is also very sensitive to *Denial-of-Service* (DoS) attacks, much more than other TCP-based protocols. This is due to the fact that, by specification, all the routes learned from a peering session are withdrawn as soon as the associated TCP connection is lost. Therefore, the ability for an attacker to reset a TCP connection or overload/reboot a router can have a significant effect. Moreover, if the connection is being reset multiple times, the *Route Flap Dampening* (RFD) protection mechanism of BGP will suppress the flapping routes for a prolonged length of time after the end of the attack.

A TCP Reset attack consists in sending a TCP packet with the *RST* flag set to one of the connection endpoints. Successfully performing this attack requires guessing the source port number of the TCP connection (which is generally non-randomly chosen), but does not require an exact guess of the sequence number. In fact, as shown by Watson [23], with an outgoing bandwidth of 1.4 Mbps it is possible to perform a successful TCP connection reset in about 10 seconds against a target with a 63 KB TCP window size.

TCP MD5 signatures [9] were introduced to protect against TCP hijacks and resets, and started being adopted by some ASes in the years 2003-2004. Still, as discovered by Convery and Franz [6], some implementations were performing the MD5 signature validation before checking for the correctness of other TCP fields (e.g., sequence number and source IP address), therefore significantly affecting the CPU usage in case of TCP DoS flooding attack. Moreover, in the same study, they showed that an attacker with the ability to sniff MD5 signed TCP packets could recover a 6 character password in 3.5 hours.

Finally, the last two more generic channels for attack should be considered: if the router accepts BGP connections (even if it later declines the peering request), DoS by resource exhaustion could be an option for an attacker. Alternatively, malformed packets could be used to attempt to crash the router or even exploit it to gain some kind of control over it. Nevertheless, these types of vulnerabilities are rare, and they are very hard to exploit.

4   Scanning the Internet for BGP Routers

The goal of our experiment was to scan the whole Internet for open BGP ports (179) and gather information about how many BGP routers are reachable and how they react to our probes. Our question was: "Is it possible for an attacker who does not already have control over a legitimate, trusted BGP router to affect the BGP infrastructure?"

Scanning the whole Internet address space is an activity that poses some challenges by itself. The number of IP addresses to probe is about 3.7 billions, which means that in order to complete the scan in 2 weeks, with maximum 1 retry per IP address, we need a constant outgoing packet rate of 6,000 SYN packets per second.

Such activity is very noisy, both from the side of the scanner's ISP and the side of the ASes receiving the scan probes. In particular, the port 179 is notoriously a very low traffic port, and, independently from how slowly the scan is performed, a whole network sweep on that port is going to be easily detected. Moreover, scanning activity is usually identified as malicious, and sometimes finds a hostile response from network administrators.

We took a number of precautions in order to reduce the impact of our scanning activity on remote networks, and we tried to address in advance potential complaints about our scanning. First of all, we setup a web page served by an HTTP server reachable at the IP address originating the scan. This web page was explaining who we were, what we were doing, and making it clear that no malicious activity was to be performed by means of our scan.

Moreover, we offered the scanned party the opportunity to contact us and request to be excluded from the rest of our scans.

Second, we minimized the incoming packet rate seen from target networks by spreading the scan of the whole network across the whole duration of the experiment. In other words, we cycled over the IP address space by incrementing the most significant bytes (network-bytes) first.

During our scanning activity we received complaints from 10 different institutions, asking to be excluded from further scans. Only one other institution (AT&T) contacted us letting us know that they were aware of our activity (see Fig. 2) and that they were interested in hearing about the details of our experiment.

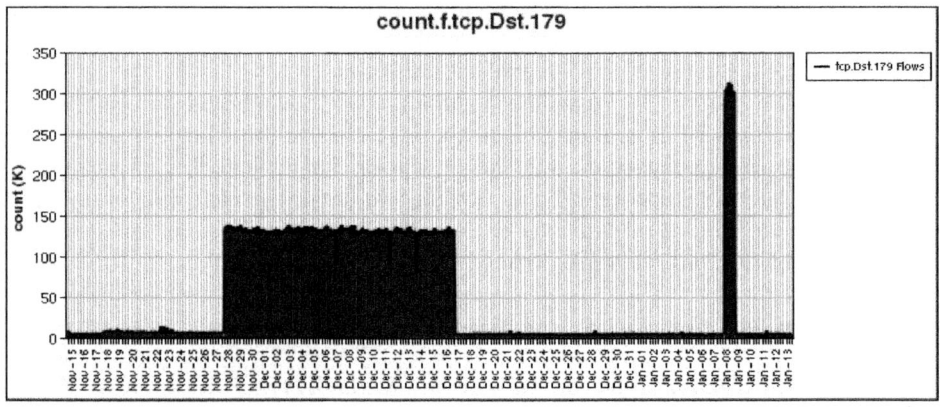

Fig. 2. Numeber of TCP flows per hour as detected by AT&T on their network (/8 prefix). The picture clearly marks two phases of our scanning activity at two different outgoing packet rates. It it noteworthy how scanning on port 179 is extremely more noisy that normal activity on that port.

One problem we incurred in was that, because of the restructuring of our lab network, we needed to change the source IP address used during the experiment. Such change was perceived by some network administrators as an evasion attempt. These administrators had actually noticed our activity and silently added a firewall rule to block our original IP address. This issue caused our IT department to request the termination of the scanning activity. The data we collected up to that point was about 73% of the whole address space, from X.X.0.0 to X.X.186.249.

Given the noisiness of such an experiment and the perceived blurred boundary between benign and malicious scanning, such a large-scale scanning constitutes a one-shot experiment: it is not viable to repeat it, at least in the near future.

4.1   Scanning Tools

We first looked at the *Scapy* [1] tool, which allows great flexibility for manipulating network packets. However, such tool does not allow handling an outgoing packet rate as high as we needed (6,000 packets per second), because of its excessively slow data handling.

We therefore decided to use *Nmap* [12] for the Internet-wide scan of hosts. Nmap was able to handle the required rate on a quad-core Intel Xeon X3220 2.4 GHz with 4GB of RAM and 1 Gbps connection to the Internet. With Nmap, we generated a report of which hosts had the TCP port 179 open.

Even though we did not have major issues, with some hindsight, we now think that implementing an *ad hoc* tool for large-scale SYN scanning would have provided us with more accurate results. Nmap has an adaptive mechanism to adjust scanning parameters, which tries to learn network delays and congestion from previous probes. While sweeping a broad range of networks, however, adaptive learning does not work well, causing unexpected (and incorrect) timing estimation (such as network RTT and optimal inter-packet sending time).

In addition, the parallel scanning of hosts performed by Nmap was not optimal. The ideal technique would have been sending out SYN packets at a chosen rate while recording possible answers. Allowing a timeout of maximum 1.5 seconds, with at most 1 retry requires one to keep in memory the state for up to 9,000 IP addresses at the same time. Unfortunately Nmap did not allow keeping a sliding window running over all the range of IPs, and, instead, it required scanning the hosts in batches. Therefore, we set to the batch size to 15,000 IP addresses, therefore alternating peaks of high scanning rate and moment of limited outgoing traffic.

5   Probing BGP Routers

Once we collected the list of IP addresses answering to our SYN packets, we tried to engage in a BGP exchange with them. Given the much smaller number of hosts to contact (about 2.2 millions), it was possible to write a Python tool for that purpose. We based out tool on the *Announcer* application by Colitti [5]. *Announcer* is an implementation of a basic BGP speaker written in Python, without any actual routing feature. We borrowed the code for generating and parsing BGP messages, while we wrote our own BGP state machine.

We took an approach that was as passive as possible. In more details, we did the following:

1. opened a TCP connection toward port 179;
2. waited for the remote peer to send any data (5 seconds timeout);
3. in case of no data received or *OPEN* request received, we sent a BGP *OPEN* request, using the UCSB AS number (AS131) and our IP address as BGP identifier;
4. waited for the remote peer to send any data (i.e., *OPEN* message, if not already sent, or *UPDATE* messages);
5. sent a *KEEPALIVE* message and waited for reply;
6. terminated the peering session.

At all times, we monitored and recorded any event that would cause an unexpected session termination:

- *timeout*, either during connection establishment or waiting for data from the remote peer (the timeout was set to 5 seconds);
- TCP *RST* packet, indicating a connection refusal or abrupt connection termination;
- TCP *FIN* packet, indicating the intention of the peer to close the connection;
- ICMP error, indicating unreachability of the remote host or port;
- BGP *NOTIFICATION* message, indicating an error during the establishment of the peering session;
- non-BGP data read from the socket, indicating the presence of a non-BGP speaker on that port.

6   Results

During our initial TCP SYN scan (carried on between December 2008 and January 2009) about 2.2 million hosts where reported to have the TCP port 179 open (i.e., 0.8‰ of the scanned IP addresses).

This pool of potential BGP routers was than probed with our Python BGP speaker in February 2009. Quite unexpectedly, 35% of these hosts no longer responded to our BGP connection requests. A reason for this behavior can be inferred by looking at the distribution of the IP addresses that were leading to timeout in the second scan: 90% of such IP addresses were belonging to less than 1% of /16 networks in the scanned address space. In fact, we found a surprisingly large number of subnets that contained a large number of hosts accepting TCP connection on port 179. The initial Nmap scan was hitting these networks at a relatively low rate, as the whole Internet address space was being covered breadth-first. When we later processed the IP addresses with the open port, these peculiar networks started receiving a much higher rate of incoming SYN packets (up to 50 packets per second). The likely explanation for the increased number of observed timeouts is the triggering of rate-limiting mechanisms in the target networks. Such networks are probably honeypots, employed to monitor malicious activity on the Internet.

The same BGP probes on the same list of IP addresses was also repeated one year later, in February 2010. As could be expected, the number of hosts timing out at our connection attempts increased by 34%. Among the IP address that resulted in a timeout in February 2009, 85% were also present in the latest scan.

Similarly to the variation in number of hosts timing out, we also observed a doubling of the hosts no longer reachable or refusing the TCP connection between the two BGP scans. Among the hosts concluding the 3-way handshake, though, the distribution of observed behaviors remained pretty much the same. Therefore, we will concentrate the results of the second scanning experiment (Table 1).

Table 1. Results of February 2010 scan. The first column indicates the last packet we sent during our probe, the second column indicates the last packet received from the remote host (or the firing of a timeout in case no data was received). The fourth column indicates the percentage over total probed IP addresses, while in the last column the percentage is computed over the hosts that completed the TCP 3-way handshake.

After we sent	event/response	number of IPs		% of probed	% of conn.
SYN	timeout	1026798		47.40	-
	TCP RST	55223		2.55	-
	UNREACH	142257		6.57	-
SYN+ACK	TCP RST	24484		1.13	2.60
	TCP FIN	402381		18.57	42.71
	BGP NOTIF CEASE	8787		0.41	0.93
	subcode 0		8559		
	subcode 5		228		
	non-BGP data	297		0.01	0.03
BGP OPEN	timeout	441130		20.36	46.83
	TCP FIN	51631		2.38	5.48
	BGP NOTIF OPEN	7843		0.36	0.83
	subcode 2		973		
	subcode 3		29		
	subcode 4		1444		
	subcode 5		5391		
	subcode 7		6		
	BGP OPEN & UPDATE	5		<0.01	<0.01
	non-BGP data	4994		0.23	0.53

Of the host that completed the TCP 3-way handshake, 45% closed or reset the connection right away. This behavior is probably determined by the interaction between a user space process and the router kernel for the management of the TCP connection. In fact, if the BGP speaker code is running as a user process, the 3-way handshake is managed by the kernel, which is not aware of the IP addresses of the configured peers (unless some ad-hoc firewall rules have been setup). The actual decision on whether the host requesting a TCP connection has the right to do so is made in the user-space process, which is notified of the connection after the 3-way handshake. Only at this point the BGP process can decide to terminate the connection.

About 1% of the hosts immediately requested us to close the BGP connection with a BGP *CEASE NOTIFICATION*, while 0.5% started communicating with our tool sending non-BGP data. Interesting enough, about 4900 hosts answered with the signature of a popular Instant Messaging (IM) protocol. Similarly to the network behavior of honeypots, such IM servers where listening on all IP addresses belonging to multiple /16 subnets on all TCP ports. Other popular non-BGP speakers we found were SSH server or telnet-like services.

0.4% (7,904) of the hosts concluding the 3-way handshake parsed our *OPEN* message, but declined the BGP peering session with a *NOTIFICATION OPEN Error* message. Almost all these BGP speakers also sent us an *OPEN* message, therefore identifying themselves. They turned out to be actually only 1258 distinct routers (i.e., distinct BGP identifiers), belonging to 318 different ASes. Among them, 3497 routers declared themselves to be in AS numbers reserved for private use, lowering to 192 the number of public ASes for which BGP routers were discovered.

In addition, we found 5 routers (all of them from the same AS) who accepted our BGP peering request and started sending us *UPDATE* messages with BGP prefixes. At this point, theoretically, we should have been able to send prefix *UPDATE*s back to the BGP router. If malicious and accepted by the remote peer, such injected route information could be used to disrupt local connectivity or backhole a subnet, potentially world-wide (i.e., advertising a subnet with a non-routable address as *NEXT_HOP*). We decided not to attempt to send such an update, as that would have been an excessively aggressive experiment. We instead contacted the administrators of such routers, in order to report the potential security threat and to try to gather some information about those BGP speakers. It turned out that we were not actually dealing with some BGP routers, rather than with a *BGPDNS* [16] infrastructure. While those hosts were not propagating incoming BGP prefixes to other BGP routers, they were storing them in a database. The owner of those BGP speakers confirmed that their security had to be improved and sending them updates could have been a way to overload their database server.

For the sake of completeness, in our scan in February 2009, we found about other 400 peers that were apparently willing to initiate a BGP peering session with us. After some investigation, these hosts appeared to be all from the same network and were not actually BGP speakers. Rather, they were behaving as *reflectors*. More in detail, they were sending us back whatever data we were writing on the TCP connection. Similarly to honeypots, reflectors are used to capture and observe malicious behavior on the Internet.

6.1 Filtering Potential Honeypot Networks

As mentioned in the previous section, we found a surprisingly relevant number of networks with a high number of IP addresses answering to our SYN packets on port 179. Such behavior is an indication that most probably such TCP endpoints are not actually BGP speakers, rather they are likely honeypot networks, whose purpose is to monitor network activity on the Internet.

Based on this rationale, we tried to identify all the networks that had more than 32 hosts answering with SYN+ACK on port 179. 2.6% of all the networks[2] satisfied our criterion, covering 81% of all the probed IP addresses. We then excluded such IP addresses from the statistics on the scan of February 2010 and compared the results.

As we expected, the percentage of IP addresses being unreachable or leading to a timeout reduced to approximately half, while the percentage of hosts closing the connection right after the 3-way handshake almost tripled (59% of all non-filtered IP-addresses). These numbers seem to confirm the hypothesis we did at the beginning of the previous section about our BGP probes hitting some rate-limiting firewalls of these honeypot networks.

6.2   Areas of Attack

From the results of the experiments we described, we can identify some areas that can be the target of attacks from a malicious user who does not have control of a trusted BGP router. First, the router that engages in a BGP peering session with an arbitrary BGP speaker is the most exposed, possibly to route injection, but also to mishandling of malformed packets. In fact, Convery and Franz [6] were able to find 4 flaws on 3 different router vendors that allowed a BGP speaker peering with a target router to force a reset of the target router.

From a more general point of view, whenever unnecessary input is parsed or resources are allocated, there is some ground for attack. For example, all routers waiting and replying to *OPEN* messages might be vulnerable to some kind of malformed *OPEN* message, or to resource exhaustion attacks.

6.3   Areas of Defense

BGP is not a public service. Every network administrator knows which ASes their routers will be talking to, and the configuration of such peers is done manually. For this reason, protection against unwanted peers is quite simple. Denial of TCP connections should be performed as early as possible, during the 3-way handshake, with cooperation between the BGP process and the kernel and/or firewall.

Moreover, the standard mechanisms for BGP protection at the TCP level should be employed, i.e., TCP MD5 signatures [9] and the TTL Security Mechanism [7]. These features should be a requirement for BGP peers rather than an option.

7   Related Work

The security of BGP and Internet routing has been addressed by multiple researchers in previous literature.

---

[2] We used the "network" separation as provided by the IP2Location (http://www.ip2location.com) database. The actual number of networks (4510) satisfying our creterion is not relevant as an absolute number, as single big networks are sometimes split into smaller ones.

A survey of possible attacks against BGP has been made in 2004 by Nordström and Dovrolis [15], in order to raise awareness of potential security risks, and to observe how existing and proposed countermeasures would have not been actually deployable or would have not been effective enough.

In 2008, Butler et al. [3] provided an updated and extensive snapshot of BGP security issues and possible solutions. They observed how BGP has been so far successful in providing a good stability of Internet routing. Nevertheless, they underlined some areas where BGP is still believed to be in need of security improvements.

Most of the literature has focused in proposing solution to the attacks where the malicious user owned a trusted BGP router. One of these proposals is *S-BGP* ([11]), which adds multiple layers of security to the current BGP specifications. S-BGP allows a BGP router to validate the authenticity and integrity of each path received from a peer. S-BGP, in fact, requires BGP speakers to sign every *UPDATE* message. The ownership of a prefix by an AS, instead, is signed by the address assignment authorities. TCP/IP level security is guaranteed by requiring the use of IPSec. Adoption of S-BGP does not appear to be feasible, as it does not allow for incremental deployment over exiting BGP, and requires many other infrastructure modifications, like the introduction of a *Public-Key Infrastructure* (PKI) and IPSec support.

Another mechanism for route validation has been proposed by Goodell et al. [8] under the name of *Interdomain Route Validation* (IRV). This proposal, rather than a modification of BGP, is an external service that allows a BGP speaker to query other ASes and the address assignment authorities in order to validate the received paths. This approach requires a much lighter deployment effort than S-BGP, but poses some concerns of scalability and vulnerability to DoS attacks.

James ([10], *soBGP*) proposed to increase the security of BGP by adding an additional *SECURITY* BGP message which carries signed information about prefixes ownership and ASes connectivity, allowing for the creation of a distributed database which routers can use to establish validity of paths. Even though soBGP allows for incremental deployment, it is not being adopted because of the necessary changes in the router software and the increased requirements in terms of router memory and CPU time.

Chan et al. [4] developed an interesting study about the adoptability of secure BGP enhancements.

In their study about behavior of spammers with respect to network resources usage, Ramachandran and Feamster [18] observed how some spammers make use of BGP prefix stealing in order to acquire control of not-yet-blacklisted IP addresses. Unused IP addresses are stolen for a short amount of time (often less than 24 hours), in order to prevent the execution of effective countermeasures. Further analysis on prefix hijacking has been carried by McArthur and Guirguis [14].

As already described, the main concerns for BGP from the DoS attack point of view, are the consequences of BGP peering session teardown. Sriram et al. [21]

conducted a simulation study on the impact of such attacks on routing performance. They showed how the BGP routing infrastructure is non-negligibly affected by this kind of attacks, even while having a relatively low success ratio for BGP peering session attacks.

A very interesting study on BGP security has been performed by Convery and Franz [6]. Their work started from an analysis and categorization of all possible vectors of attack against the BGP protocol. As the second step, they tried to discover potential vulnerability in routers by means of protocol fuzzing and resource exhaustion attacks. In the end, they performed a scanning study very similar to the one we performed. They ran a traceroute towards each prefix advertised in the Internet routing tables in order to gather IP addresses of potential BGP routers. Afterwards, they tried to connect to all of these hosts on port 179, and they sent an *OPEN* message and listened for data. Our approach for collecting IP addresses of routers was performed on a much larger scale. In fact, we were able to conclude the 3-way handshake with 2.2 million hosts on port 179 (vs. 4,602). Clearly our number is inflated by the fact that we hit not only routers, but also some honeypot/reflector networks. However we were able to reach a broader number of BGP speakers, not only those we could encounter on a path from our network towards all ASes on the Internet. In fact, in February 2009, about 16,000 hosts answered our OPEN request (vs. 1,750), and we found that 7 of them were willing to establish a BGP peering relationship with us.

8   Conclusions

With our experiment we took a snapshot of a very large subset of the Internet, looking at how BGP routers answer to a casual inquirer.

We described the challenges we faced for performing such large-scale scanning experiment, and what we learned. Such activity revealed itself to be very noisy and perceived as extremely hostile. Therefore, one performing a large-scale scan must be able to handle complaints, no-scan requests, and be aware that she may not be able to repeat the experiment.

Based on how the BGP protocol works and on related work on the subject, we identified some possible areas of attack to the BGP infrastructure. We then tried to match them to the results of our scan.

We could not find any real immediate threat. Nevertheless, we found a large number of BGP speakers that would establish a connection with us and possibly accept some input from us. Such routers were not required to engage in such communication, and, in doing so, they exposed themselves to a potential menace, as BGP is extremely sensitive to Denial-of-Service attacks. Moreover, given the amount of trust given to the other peers, a compromised router could seriously affect the whole Internet routing infrastructure. For this reason, it is very important that some equivalent of the principle of *least privilege* is applied also to the acceptance of BGP connections, shielding the routers as much as possible from potentially malicious traffic.

Acknowledgments. This work has been partially supported by the National Science Foundation through grants 0905537 and 0716095, and by the Office of Naval Research through grant ONR N000140911042. We would also like to thank AT&T for their feedback on our scanning activity.

References

1. Biondi, P.: Scapy (2009), http://www.secdev.org/projects/scapy/
2. Bono, V.J.: 7007 Explanation and Apology (April 1997), http://www.merit.edu/mail.archives/nanog/1997-04/msg00444.html
3. Butler, K., Farley, T., McDaniel, P., Rexford, J.: A survey of BGP security issues and solutions. AT&T Labs Research (2008)
4. Chan, H., Dash, D., Perrig, A., Zhang, H.: Modeling adoptability of secure BGP protocol. ACM SIGCOMM Computer Communication Review 36(4), 290 (2006)
5. Colitti, L.: Active BGP Probing (2009), http://www.dia.uniroma3.it/~compunet/bgp-probing/
6. Convery, S., Franz, M.: BGP Vulnerability Testing: Separating Fact from FUD. In: Black Hat US 2003 / NANOG28 Meeting (2003)
7. Gill, V., Heasley, J., Meyer, D.: The Generalized TTL Security Mechanism (GTSM). RFC 3682 (Experimental) (February 2004), http://www.ietf.org/rfc/rfc3682.txt; obsoleted by RFC 5082
8. Goodell, G., Aiello, W., Griffin, T., Ioannidis, J., McDaniel, P., Rubin, A.: Working around BGP: An incremental approach to improving security and accuracy of interdomain routing. In: Proc. NDSS, vol. 3 (2003)
9. Heffernan, A.: Protection of BGP Sessions via the TCP MD5 Signature Option. RFC 2385 (Proposed Standard) (August 1998), http://www.ietf.org/rfc/rfc2385.txt
10. James, N.: Extensions to BGP to support secure origin BGP (sobgp). Network Working Group, Cisco Systems (2002)
11. Kent, S., Lynn, C., Seo, K.: Design and analysis of the secure border gateway protocol (S-BGP). In: Proc. of DISCEX 2000 (2000)
12. Lyon, G.: Nmap – Free Security Scanner For Network Exploration & Security Audits (2009), http://www.nmap.org
13. Mahajan, R., Wetherall, D., Anderson, T.: Understanding BGP misconfiguration. In: Proceedings of the 2002 Conference on Applications, Technologies, Architectures, and Protocols for Computer Communications, pp. 3–16. ACM, New York (2002)
14. McArthur, C., Guirguis, M.: Stealthy IP Prefix Hijacking: Dont Bite Off More Than You Can Chew. In: Proc. ACM SIGCOMM (2008)
15. Nordström, O., Dovrolis, C.: Beware of BGP attacks. ACM SIGCOMM Computer Communication Review 34(2), 1–8 (2004)
16. Oppermann, A., Jeker, C.: BGPDNS, Using BGP topology information for DNS RR sorting a scalable way of multi-homing. RIPE 41 Meeting (2002)
17. Pilosov, A., Kapela, T.: Stealing The Internet. DefCon 16 (2009)
18. Ramachandran, A., Feamster, N.: Understanding the network-level behavior of spammers. ACM SIGCOMM Computer Communication Review 36(4), 302 (2006)
19. Rekhter, Y., Li, T., Hares, S.: A Border Gateway Protocol 4 (BGP-4). RFC 4271 (Draft Standard) (January 2006), http://www.ietf.org/rfc/rfc4271.txt

20. RIPE NCC: YouTube Hijacking: A RIPE NCC RIS case study (2008),
    http://www.ripe.net/news/study-youtube-hijacking.html
21. Sriram, K., Montgomery, D., Borchert, O., Kim, O., Kuhn, D., et al.: Study of
    BGP Peering Session Attacks and Their Impacts on Routing Performance. IEEE
    Journal on Selected Areas in Communications 24(10), 1901 (2006)
22. Villamizar, C., Chandra, R., Govindan, R.: BGP Route Flap Damping. RFC 2439
    (Proposed Standard) (November 1998), http://www.ietf.org/rfc/rfc2439.txt
23. Watson, P.: Slipping in the Window: TCP Reset attacks (2004)

Securing the Core University Business Processes

Veliko Ivanov, Monika Tzaneva, Alexandra Murdjeva,
and Valentin Kisimov

University of National and World Economy,
8 December Str., Student Town, 1700 Sofia, Bulgaria
veltex@abv.bg, monika_tzaneva@yahoo.com, amurjeva@abv.bg,
vkisimov@gmail.com

Abstract. In the paper are presented solutions for securing the core University Business Processes. A Method for identification which Business processes are critical for security point of view, on which is required to pay more attention for its securing. For securing of the elected Business processes is developed a new security system – Extended Certification Authority. Special Secure eDocument Management Architecture is developed, on which base are developed the solutions for securing the following University Business processes - Delegation of exam permissions to lecturers, Recording exam marks, and Exchange management documents.

Keywords: Securing business process, Extended Certification Authority, Secure eDucument Management Architecture.

1 Introduction

There are few definitions of the term "Business process", which are not too different each other. In our research we have accepted the definition of [4], which under Business process understands a set of interrelated tasks leading to create a product or a service. Each Business process needs to have required security of its execution. This means the process should provide the security triangle of parameters Availability, Confidentiality, and Integrity, in an appropriate level.

University is like any other enterprise – with a set of business processes, from which some are more critical from security point of view. The criticality is coming from the generation of the university end result – the degree document, where all exam marks make its content. The processes leading to forming the university end result defines the set of core processes. Based on the mentioned criteria for the criticality, it is not too difficult to list the university core business processes.

The main goal of the current research is to identify what level of security the core business processes need, e.g. which the core university business processes needing higher level of security are, what is the needed security level, and how to provide that security.

J. Camenisch, V. Kisimov, and M. Dubovitskaya (Eds.): iNetSec 2010, LNCS 6555, pp. 104–116, 2011.

2   Method for Selection of Core Business Processes Needing High Security

There is not standard way for selection of core business processes which need some level of security. There are some practical approaches - [5] and [6], which resolve particular private needs, but no one exists for university business processes security evaluation. For this reason we have developed our "Method for selection of core business processes needing high security".

The proposed Method works with number of University Business processes, which core set consists of 11 processes, identified in general, and more specifically for the case for our particular research – University of National and World Economy (UNWE), Sofia, Bulgaria. These 11 University Business Processes are:

- Delegation of exam permissions;
- Recoding exam marks;
- Exchange Academic Counsel (AC) reports;
- Exchange Rectors' Counsel (RC) reports;
- Exchange Faculty Board reports;
- Exchange Departmental Meeting reports;
- Review Public statements;
- Change the Educational Curriculums;
- Admittance of new students;
- Admittance students from 2nd to 3rd year;
- Change student status.

These business processes are evaluated from the needed security point of view via 9 criteria, grouped in 5 groups. The evaluation is provided with metrics 1 to 3, where 1 means needed low security requirement, 2 means medium and 3 means high. The following criteria are identified with the specified value and security impact:

a) Business Continuity – the criterion identifies what are the requirements for the Business continuity from the University business process;
b) Disaster Recovery – the criterion identifies the need for available recovery and the time for recovery of the University business process;
c) Event management – the criterion identifies how critical are the events for security violation for the business process;
d) Availability – the criterion identifies how critical is the business process to be available. Normally the availability is measured in percentage, but for the purpose of the current Method, these percentages are transformed into value 1 to 3 (from 0 to 33% - value 1, from 33% to 66% - value 2, and up to 99% - value 3);
e) Confidentiality – the criterion identifies how critical is to have encryption of the information for the business process and the level of encryption. This is an integrated criterion, where the percentage of the encrypted messages is evaluated, the type of encryption (symmetrical / asymmetrical), the key length, and the period of key refreshment;

f) Integrity – the criterion identifies the need of information integrity provided via security means;

g) Business impact – the criterion identifies what is the Business impact of the Business process to the entire University business security. This criteria is formed from a few sub-criteria:

- Relative number of beneficiaries (students, professors, staff) impacted from the Business process;
- Relative number of providers (external and internal) impacted from the Business process;
- Level of cost of this Business process compared with the total cost of university processes;
- Level of monthly transactions executed from the Business process;
- Political sensibility of the Business process, which relates to the view of the University to the country and to the entire world, the effect of the Business process to the reputation of the University, the effect of the Business process to the high-education process in Bulgaria and Europe;
- What is the overall impact to the University from the specific Business process;
- What is the overall impact to professors (lecturers) and students from the specific Business process;

The collection of values from the above mentioned sub-criteria for Business impact is proposed to be done via a table – Table 1:

Table 1. Forming of Business Impact criteria

Business processes	Relative number of benefi-ciaries impacted	Relative number of providers impacted	Level of costs of this business process compared with total cost of university	Level of of monthly transac-tions	Political sensitivity	What's the overall impact on the University	What's the impact on professors	What's the impact on students	*Total average score*
BP #1									
BP #2									
etc.									

Each sub-criterion is also evaluated via the measures from 1 (low level) to 3 (high level). For each Business process a Total average score is created, which is rounded to the values 1, 2 and 3. This value is the defined value for Business Impact for the specific Business process.

h) Risk level – the criterion identifies the level of which the risk of the entire University business can raise, based on the current business process. The University Business strategy and University Security strategy define levels of risk (there are no many universities in the world which have developed Business strategy and Security strategy, like the big corporation. In the current case of UNWE, the research team has identified the appropriate risks, based on interviews with the university management);

i) Minimum acceptance level – the criterion identifies which is the minimum security level, which is needed for the specified Business process. This criteria is formed from a few sub-criteria:
  ▪ Eligibility - inquires and responses of the Business process;
  ▪ Enrollment and Disenrollment of data for/to the Business process;
  ▪ Authorization – requests and responses to the Business process;
  ▪ Claims – receipts and adjudications from the execution of the Business process;
  ▪ Claims status - requests and responses from the execution of the Business process.

The collection of values from the above mentioned sub-criteria for Minimum acceptance level is proposed to be done via a table – Table 2:

Table 2. Forming of Minimum acceptance level criteria

Business process	Eligibility – inquiries & responses	Enrolments & Disenrollments	Authorization – requests & responses	Claims – receipts & adjudications	Claims Status – inquiries & responses	*Total average score*
BP #1						
BP #2						
etc.						

Each sub-criterion is also evaluated via the measures from 1 (low level) to 3 (high level). For each Business process a Total average score is created, which is rounded to the values 1, 2 and 3. This value is the defined value for Minimum acceptance level for the specific Business process.

The proposed Method for selection of business processes and their level of needed security uses a table, shown in Table 3 below.

Table 3. Business processes selection for needed security

#	Business process	Participating in:			Degrade level			Bus. impact	Risk level	Min Accept. Level	Total score
		Bus. Cont.	Dis. Recov	Event Mgmt	Avai- labil.	Confi- dent.	Inte- grity				
1	Delegation of exam permissions	2	2	2	2	3	2	3	1	3	20
2	Recoding exam marks	3	3	2	3	3	3	3	2	3	25
3	Exchange AC reports	3	3	1	2	1	1	3	1	3	18
4	Exchange RC reports	3	2	1	2	1	1	3	1	3	17
5	Exchange Faculty Board reports	2	1	1	1	1	1	2	1	2	12
6	Exchange Departm.Meet. reports	2	1	1	1	1	1	1	1	1	10
7	Review Public statements	3	1	1	3	1	1	3	2	3	18
8	Change the Educ. Curriculums	1	1	1	1	1	1	1	1	1	9
9	Admittance of new students	3	3	2	3	1	1	3	2	2	20
10	Admittance students 2nd to 3rg year	3	2	2	3	1	1	2	1	1	16
11	Change student status	2	2	1	2	2	3	1	1	1	15

For the target object – UNWE, particular values are resulted into the table. From those results we have concluded, that the core university business processes from security point of view are:

- **Delegation of exam permissions (Protocols) to lecturers;**
- **Recording exam marks; and**
- **Exchange management documents**
  - Exchange AC reports;
  - Exchange RC reports;

The Business process "Admittance of new students" has also relatively high score – 20, but because it is with relatively small infrastructure part of the University IS infrastructure, we will exclude it from our research activities. The Business processes Exchange Academic Counsel (AC) reports and Exchange Rectors' Counsel (RC) reports we can combine for the future presentation of research results, because their security requirements are closed. In this way we receive an aggregated business process called "Exchange management documents". For this reason, the mentioned 3 core business processes will continue to be the focus of the further research, presented in the paper.

3   Extended Certification Authority

For securing the core university business process is required a special solution with involvement of Certification Authority (CA) and Public Key Infrastructure (PKI). The required functions for securing the core business processes are logically between the functions of CA and functions of PKI.

Certification Authority generally is either independent system or it is an agent of a PKI. The user can do the main certification functions through its keys and certificate – authentication, encryption, digital signature, data integrity, non-repudiation, etc., while the certificate can be used also for reputable identification, timeframe for validity and specification of possible security functions. The security credentials are linked to the CA via the digital signature of the certificate, where the Public key is included, and the Public key recognizes the pair connection with its Private key. CA is an entity that issues security keys – Private and Public and Digital Certificates for use by other parties. The functions of the CA can be summarized into:

- Has its own Root certificate;
- Verify the identity of entities asking to issue certificate;
- Generate Private and Public keys;
- Issue digital certificate attesting to the identity;
- Digitally sign the certificate via its Root certificate;
- Store certificate and keys in the secure tokens;
- Play a role for the trust party;
- Maintain Certification revocation (CR) process, CR List and Repository
- Use OCSP protocol (On-line Certificate Status Protocol) for access to CR List

From other side, the PKI is set of hardware, software, policies and procedures needed to issue and maintain Asymmetrical and Symmetrical keys and Public Key Certificates incorporating different user's identities, and also to order, issue, register, store, distribute, renew, revoke and manage Public Key Certificates. PKI serves as a trusted third party between many end-users. The functions of the PKI are much bigger and complex than the CA, and they can be summarized into:

- Operate with CA, Registration Authority (to verify and accept requests for certificates) and Repository (repository for certificates and CR List);
- Provide Backup and Recovery for the purpose to restore lost or damaged Certificates;
- Update Key History – at any certificate change, to update the history logs (because of expiration or a name change). Any data secured using the older keys would not be accessible unless the older keys are kept in an archive;
- Revoke Certificates, when the Certificate is no anymore valid or it is discredited;
- Automatic Key Renewing and Certificate renewing – after the expire of the certificate, new Private key and Public key has to be issued, on which base a new Certificate has to be issued for the same user, with a process of automatic renewal of them to the end-used repository – smart cards or USB devices. Automated key recertification can update the certificate with a new expiration date when necessary, without manual intervention;
- Cross Certification – used to establish a trusted relationship between separate PKI's. This allows for a distributed and decentralized infrastructure;
- Support for Non-Repudiation – prevents a certificate owner from denying that data was sent using the owner's certificate;
- Time stamping – certifies that the time stamp on the secured data is set and it is accurate and valid;
- Client API – A means for an application to use the services offered by a PKI.

PKI is an expensive and difficult to operate system. Only limited number of companies has the luxury to own a PKI. University generally does not operate with budgets of the big companies, for which reason it is not practical to expect that a University will create or buy a PKI. The world has established and easier approach to the PKI, defining "Small PKI". Small PKI was born out of a joint effort to overcome the over complication and scalability problems of traditional PKI, decreasing the role of RA, Repository, and some of the functions. The functions of the Small PKI cover all functions of the CA, with incorporation of some functions of the PKI. In summary, the functions of Small PKI can be listed as:

- Full CA functions;
- Mechanisms to support security in a wide range of Internet applications, including IPSec protocol;
- Keys renewal and key management;
- Encrypt electronic mail and WWW documents;
- Range of Secure application functions requiring use of Public key Certificates as ePayment and B2B;
- Support a range of trust models.

The analysis of the security needs for securing the core university business processes shows that the functionality of the Small PKI is too rich from security point of view and a university does not need of such a system. Here we have to add that the Small PKI is also an expensive system for the university budget and to securing the core University Business Processes we need a system, which functions are between those of CA and Small PKI. We call such a system "Extended CA" and position it in the functional axe of crypto functionality in a way, shown in figure 1.

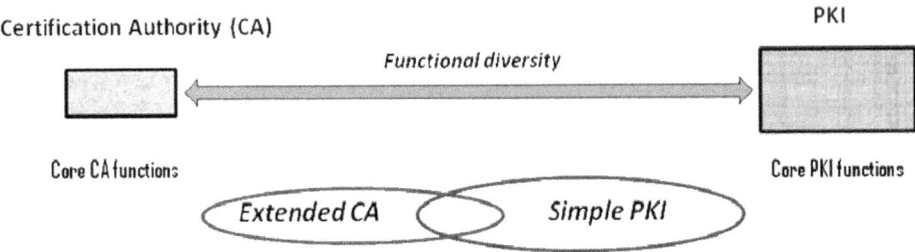

Fig. 1. An axe of crypto functional systems, where the place of "Extended CA" is defined to be between CA and Small PKI

The authors have analyzed the functions needed for securing the core university business processes, providing cost-functions analysis. On this base a crypto system with the name "Extended CA" was designed, which is relatively cheap from university budget point of view and functionally reach for security point of view. We have found that the Extended CA has to be a CA with LDAP, Web server for communications, and supporting some PKI services such as digital signature, keys exchange, email encryption, and secure token support. After that analysis, we have concluded that the Extended CA has to have the following functions:

- To function of Full Certificate Authority (CA) and support of Certificate Revocation Lists (CRL);
- To operate with Directory server for storing security credentials;
- To use Web server for users' communication, exchange of information and auditing of all security transactions and operations;
- To provide Kerberos authentication;
- To use tokens (Smart cards and/or USB devices with smart cards) to store the security credentials inside;
- To support Securing web servers;
- To support Securing email;
- To provide Digital signature in off-line and on-line modes;
- To execute Application signatures (i.e. Signed drivers or ActiveX controls);
- To support Encrypted File Systems (EFS) and work with appropriate recovery agents;
- To support Smart Card Logins.

There are two standards (RFC) for certificates – for Public Key Certificates [7] and for Attribute Certificates [8]. The most used certificates in the world are the Public Key Certificates. At the same time, the Attribute Certificates provides some futures, which can be used for securing in a good way the university business processes. Generally, attributes in Attribute Certificate offer user's short-lived information and dedication, such as user's roles and access permissions, which suite the university short term access permissions, like giving access of a professor to exam marks for a subject, valid for a block or semester. For this reason the capabilities of the attributes in Attribute Certificate is preferably to use for the university purposes. For this reason we have analyzed the suitability of incorporation both certificates values into a single certificate, covering the international standards. We went to a conclusion, that the Extended CA can provide certificates, which combine the usability of both certificates.

Our proposed solution for a Certificate, issued by the Extended CA is to include the necessary attributes, which normally are part of the Attribute Certificate, in the Private Extensions of the Public Key Certificate. Applying the proposed approach, the certification issued by the Extended CA will have the following value-added functionality:

- Secure identification and authentication using pure SSL protocol + LDAP server;
- Separation access control based on SSL managed certificates;
- Add complementary security credentials, e.g. second PW, Biometric identification, etc.;
- Easy to manage Private information in Corporate (in our case – University) systems;
- Public Key Certificate (PKC) is per user, the attributes are "per group", and integration of both is for relatively medium-term certificate, e.g. per semester;
- Provide "Revocation" for attributes;
- Security policy defined per medium-term (per semester);
- Provide trust from different CA/PKI via incorporated additional PKC (another way for cross certification);
- Incorporate Privilege attributes as Copy rights, Patents, Trademarks, access to management documents (levels and specific);
- Internet and Intranet specific rights;
- Role based control using attributes.

4   Secure eDocument Management Architecture

The elected core University Business Processes for security operate predominantly with electronic documents. Like in any corporate Information System, solutions for a few problems can be a single integrated architectural solution. For this reason we have developed a special Security architecture, which will be the base for the securing the elected core University Business Processes. The developed architecture we called "Secure eDocument Management Architecture". The purpose of the offered in the paper architecture is to manage in a secure different management documents, providing for them the necessary security features, which do not exist in general electronic documents. The Secure eDocument Management Architecture has two focuses of security features:

a)   Securing the elements of the electronic document (eDocument);
b)   Securing the transactions with the eDocuments.

The Secure eDocument Management architecture operates with Lecturer's PC, where the Secured document is processed, and with Web server, where the University Business processes have centralized for treatment. As part of the Web site is the explained above Extended Certification Authority. The Secured document operates with 3 security features:

- Marcos, activated on loading the document;
- eButton with integrated macros, activated on pressing the Button;
- Security fields, keeping encrypted information, which is decrypted on Lecturer's PC.

Graphically, the Secure eDocument Management Architecture can be presented in figure 2.

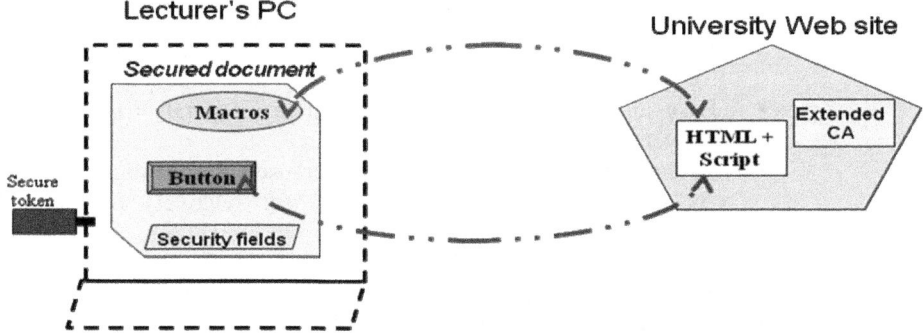

Fig. 2. Conceptual architecture of Secure eDocument Management Architecture

The mentioned 3 security features are part of the eDocument. Having at least one of those security features, the eDocument is converted to a Secure eDocument. Each one of those 3 security features can exist in many instances in the Secure eDocument. The first security feature - Macros are programs, which are executed during the transaction (transferring) of the eDocuments from the Web site to the Lecturer's PC. For example, when a document is loaded from the University Web site to a Lecturer's PC, one of a few Macros can be executed automatically. The second security feature - Buttons are also programs, but activated by the user of the eDocument. For example, when a lecturer wants to digitally sign a document, he has to press a button, which keeps virtually a program code executing the process of digital signature. The third security feature – Security fields are part of the eDocument, which can be treated with different security algorithms, for example the field can be encrypted and only with the wish of the end-user (lecturer) the field would be decrypted and represented in a plain text (via pressing of a button or executing of a Macros).

The developed Secure eDocument Management Architecture is used in securing of the presented below 3 core University Business processes.

5 *Secure Process*: Delegation of Exam Permissions to Lecturers

The exam permission is a university document called Protocol, which specify which students to which subject have rights to be examined, by which lecturer. Not all student studying appropriate subject have rights to be examined at the end of given semester – they are different university procedures permitting under which conditions a student can go to be examined, for example if he finalized all assignments during the semester.

The current process can be secured using the explained in the previous section Secure eDocument Management architecture. We have designed, the Macros embedded in the eDocument – Protocol to provide the following functions:

- Full document protection via Password (via the lecturer's ID);
- Partial Confidentiality – some fields are only readable by the specified lecturer - Protocol name, Protocol number, semester validity;
- Document data integrity – Protocol checking on receiving;
- Checking the document time period validity: Document data creation – Document date receiving;
- Document digital signature verification.

The Macros also provide 3 additional security functions, related to the entire eDocument:

- Protocol cannot be read and used by non-dedicated lecturer;
- Protocol cannot be operated by non-dedicated lecturer;
- Lecturer cannot take a Protocol, which is not created by the University Information Security system.

The eDocument-Protocol is developed with 2 buttons:

- Accepting the document after visual control;
- Non-repudiation of receiving the eDocument.

6 *Secure Process*: Recording Exam Marks

The process of Recording the exam marks is associated with moving the exam marks from the eDocument-Protocol to the Information system "Student", where all exam marks are recording. Based on those exam marks are generated the degree documents. The manual process is related to many paper records, and using many clerks. At the end, the responsibility for the exam results is to the lecturer, but many other people participate between the lecturer and the computer record of the exam marks. To eliminate all those interim people who do not keep responsibilities, is developed the current security process.

The securing of that business process uses a few security features:

- Secure eDocument-Protocol;
- Digital signing of the Protocol by the lecturer, before entering it into the system "Student";
- Using of One Time Password (OTP) for additional authentication of the lecturer;

The sequence diagram of the securing of the specified University business process is presented in figure 3.

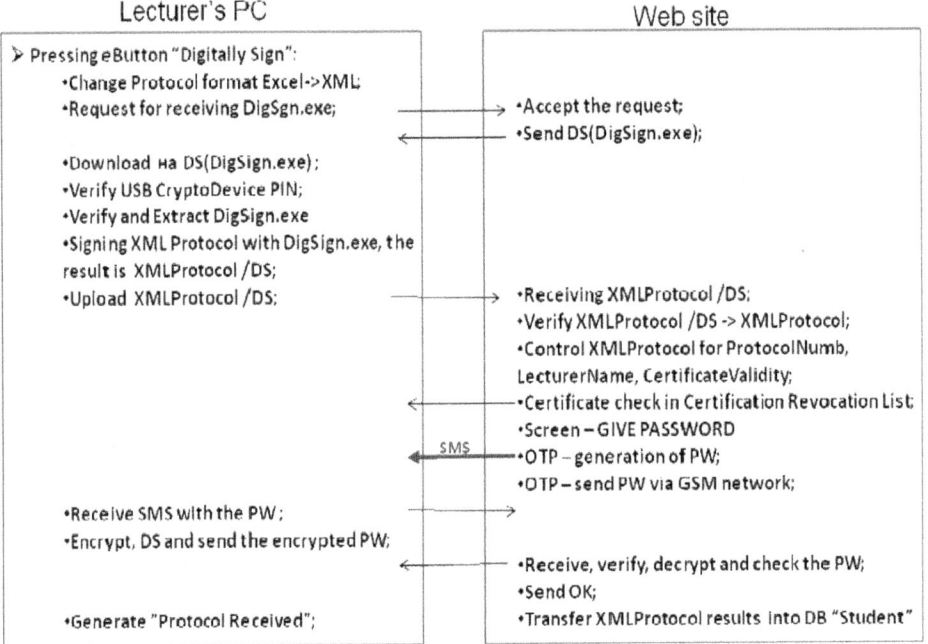

Fig. 3. Control flow of the secure technologies used to secure the Business process "Recording the exam marks"

There are few special security technologies used in that diagram:

a) The program for Digital signing of the eDocument-Protocol is always loaded from the University Web site, just before the signing. This ensure correct digital signing;

b) The lecturer uses an USB Crypto device, which stores the Private key and the Certificate of the lecturer. To access those security credentials, the lecturer should enter the PIN code of the USB Crypto device;

c) During the signature verification, the major elements of the lecturer's certificate are checked, like whether that lecturer has right to enter the Protocol in the system "Student", what are the credibility of the lecturer in its certificate according to the subject for which the exams are entering, etc.;

d) The complex signature verification includes also the check whether the certificate is part of the Certification Revocation List (CRL). In the proposed solution CRL is not web accessible, but via an internal interface, because the check is provided inside the University Web site, where the Extended CA is also part of;

e) The OTP has a special solution which ignores its major security weakness – man-in-the-middle. This solution implement encryption and digitally signing the received password by the lecturer and then send it to the Web site. These both technologies identify and authenticate exactly the lecturer and the University Web site, working with entering the exam marks.

7 *Secure Process*: **Exchange Management Documents**

The Business process of Exchange management documents is designed as Peer-to-peer exchange of encrypted documents. This peer-to-peer link is a logical link, using the University Web site as an interim and control station. The main message moving mechanism used in the proposed solution is the emailing. The idea is to use exactly two interim stations – Web site and email server. The Web site is part of the Secure eDocument Management Architecture, which use as subsequent messaging mechanism for emailing. The conceptual architecture of that security architecture is presented in figure 4.

It is important to mention that the email services which are involved in the solution can be a corporate Email server or a Cloud Computing email services.

The developed solution provides as additional security features:

- Detail logging of all transactions – in the Web server;
- Auditing of all transactions – also in the Web server;

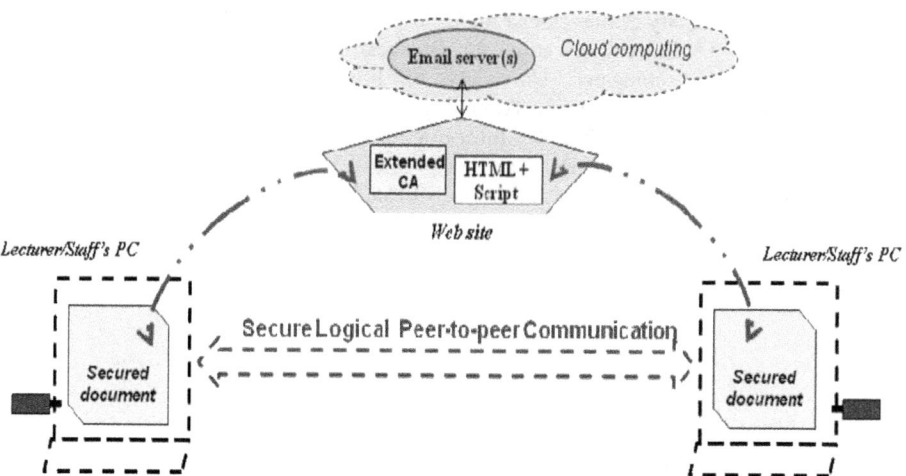

Fig. 4. Architecture for secure exchange of management documents

8   Conclusions

The presented security solutions for securing of the core University Business Processes are based on electronic document incorporating security program code in a few ways, integrating existing security technologies, but executed in a secure ways, using new security system – Extended Certification authority. The prototyping of the proposed solution uses documents based on MS Office, while the security technologies for them have been developed in two platforms – Java and VB-.Net. Both platforms showed equal flexibility for operation and security quality. The proposed research has few open research problems, like a Framework for development of Extended Certification Authority, and well securing the University Web site (with integration with Extended CA), if it is not in the University Intranet, but on open Internet.

References

1. O'Higgins, B.: What is the Difference Between a Public-Key Infrastructure and a Certification Authority?
   http://www.opengroup.org/comm/the_message/magazine/mmv4n2/pki.htm
2. Shirbu, M., Chuang, J.: Distributed Authentication in Kerberos using Public key cryptography,
   http://people.ischool.berkeley.edu/~chuang/pubs/pkda.pdf
3. Bass de Graeff, Business processes Security, Unisys,
   http://www.unisys.com/unisys/ri/wp/detail.jsp?id=17600051
4. Kisimov, V.: Dynamic Business-managed Information systems, Sofia, Bulgaria (2008) ISBN 978-954-323-444-8
5. Garfamy, R.: Supplier selection and Business process improvement, doctoral thesis, Autonomous University of Barcelona (2005)
6. Lee, J., Choi, J.: Process selection for Business Process Management in a mobile telecommunications company, University of Illinois at Urbana-Champaign, USA. International Journal of Information Technologies and Management 8(4), 382–399 (2009)
7. RFC 2459, Internet X.509 Public Key Infrastructure Certificate and CRL Profile,
   http://www.ietf.org/rfc/rfc2459.txt
8. National RFC 3281, An Internet Attribute Certificate Profile for Authorization,
   http://www.faqs.org/rfcs/rfc3281.html

Some Technologies for Information Security Protection in Weak-Controlled Computer Systems and Their Applicability for eGovernment Services Users

Anton Palazov

University of National and World Economy,
Department of Information Technologies and Communications,
UNSS, 1700 Sofia, Bulgaria
apalazov@rubella.bg

Abstract. The users of eGovernment services start exchanging documents with administrative authorities, making ePayments, and in such communications the risks of confidential information disclosure and direct financial losses are growing up. The computer systems of these users are weak-controlled and are outside of sphere of well-defined information security protection decisions. The technologies for data protection in case of theft or loss of computers and data devices and in case of data leakage are very important for eGovernment services users and must have appropriate properties to be useful for their security needs. A model of anti-theft technology implementation, which disables stolen computers and can send them data-destructive commands to erase sensitive data, is presented. The technologies for control over the channels which can lead to data leakage protect data by whitelisting or blacklisting some devices or ports, by prohibit and allow some actions and operations, or by transparent encryption of outbound data. Some technologies for control over the leaving data use pre-defined set of sensitive data type definitions. Users can select definitions to apply or can customize some of them according specific conditions or regulations. At the end some conclusions about applicability of anti-theft and sensitive data leakage prevention technologies for protection of information security of eGovernment users was done.

Keywords: eGovermnemt services users, anti-theft, data leakage prevention, sensitive data type definitions.

1 eGovernment Users Evolution and Risks for Their Information Security

In time, the users of eGovernment services get more experience and possibilities and start performing more complex tasks - exchange of documents with administrative authorities in electronic form which can have legal consequences for both sides, making ePayments for tax duties, receiving money from social funds etc.

J. Camenisch, V. Kisimov, and M. Dubovitskaya (Eds.): iNetSec 2010, LNCS 6555, pp. 117–122, 2011.
© IFIP International Federation for Information Processing 2011

In such more complex communucations the risks of confidential information disclosure and direct financial losses are growing up and information security protection system must mitigate them and counteract to the fault. Computer theft, data device losses and leakage of important user data are the events, which can transform these security risks to attacks against user privacy.

The survey of Ponemon institute [11] for endpoint security in 2009 shows that the loss of sensitive data is real event for about 50 percents of analyzed companies (respondents). The results for lost or stolen computing devices are almost the same. The interviewed security professionals give the opinion that these risks will be one of their main troubles in next 12 months. The users of eGovernment services interact with the administrative authorities mainly with their home computers thru the Internet. These computers are outside the protecion of information security based on well-defined from computer professionals, good supported and strongly enforced security policy as is the case of corporate information systems. From point of view of information security protection the computer system of typical eGovernment users are weak-controlled and in many cases there is no implementation of clear and reliable security rules [1], [2].

So, the technologies for data protection in case of theft or loss of computers and data devices and in case of data leakage are very important for eGovernment services users and must have appropriate properties to be useful for their information security protection.

In this research the following steps are performed:

- Study and analysis of publications in magazines and in Internet and defining of criterias for estimating of technologies applicability in information security protection;
- Specifying of main parameters and characteristics of analyzed technologies;
- Measuring and conclusions formulation about analyzed technologies applicability in information security protection of eGovernment services users.

2 Technologies for Client-Side Protection for Lost or Stolen Computer Systems and for Stand-Alone Data Leakage Prevention

Endpoint computer systems (servers, desktops, laptops) and removable media (CD, DVD, USB memory) are especially vulnerable to security accidents like loss or theft, which makes them a weak spot in information systems infrastructure [10]. Companies and citizens need security solutions that can protect their information systems against these threats and ensure that unauthorized persons have no access to their saved sensitive data.

2.1 Protection of Lost or Stolen Computer Systems and Data Devices

Loss or theft of computer systems, devices or data files is significant risk for information confidentiality. Security protection technologies must react on these events with registering of stolen systems, disabling them, sending a data-destructive commands to erase sensitive data and easily reactivating them if they are recovered.

The vendors of Endpoint protection platforms [5], [7] offer different solutions for data protection in case of theft or loss of computer systems. The basic characteristics of these solutions can be summarized as:

- Protected user computers have internal timer (software module or hardware device) which try to communicate with Ownership Management Server via Internet. The duration of the interval between two connections can be set from the user or remotely from the server or administrator;
- If the user computer can't connect to the Management server in the defined for the timer interval or communication schedule, the timer suppose that the computer was lost and activates the transparent service which disables user sensitive data and the computer system itself;
- If the computer was lost, its user can call the Ownership Management Center help desk and can register the event. When the lost or stolen computers connect later to the server in their regular communications, the server can send to the workstation self-destructive signal. The internal timer on the lost computer system react to that signal and disables in some ways sensitive user data and probably the computer system itself;
- The Ownership Management Server has possibilities to submit the one-time activation key, which can be used to enable the computer in case it was found and its owner can authenticate himself.

By adding this client-side intelligence the confidentiality of sensitive user data can be reliably protected in case of loss, theft or suspicious circumstances - the internal timer on client computer systems or the signal from the server will automatically disable these data and will not allow the attacker to exploit them.

2.2 Standalone Endpoint Data Leakage Prevention

The technologies for endpoint data leakage protection can be divided in 2 general groups:

- First, technologies for control over the channels, which can lead to data leakage (data devices, ports, Internet services), and;
- Second, technologies for control over the data, which leave the computer system in different directions.

Technologies for channel control [8] protects data by whitelisting or blacklisting specific devices or ports, by prohibit and allow some actions and operations in available channels, or by transparent encryption of data that go out of information security protection system. In case of encryption users can apply existing PKI infrastructure or can define their own keys or passwords for removable media, file types or specific files and communication units which will be exchanged with their partners.

Endpoint users need to encrypt some removable media to protect data confidentiality and to allow access to them via some unlock mechanism (passwords, keys for decryption, etc), which is shared with other partners. They can protect the whole media, particular file types, or specific data files. Central management helps security administrators to create, implement, enforce and audit secirity policies for different groups, users or business partners.

Technologies for control over the leaving data offer possibilities for content detection, which, by example, can include SSN or VAT number identification in exchanged documents. Some solutions [9] can detect in outbound flow other specific "registered" data elements such as database field values and aggregates, file names, register keys, etc. These tools use number of dictionaries where are defined, by example, financial items, legal term, trade marks. The dictionaries allow using of "wild card", operators and case-sensitivity indicators. Some of them have intelligence features which do more than just a formal search in data content.

Both groups of technologies provide decisions for data leak prevention by restricting unauthorized export of confidential data via certain communication ports or peripheral devices. Security policies define some read and write restrictions on ports and devices based on device type, data file types or even on individual peripheral devices or files. Additional software components apply and control compliance with defined security rules.

Some vendors [7] offer tools which include valuable set of sensitive data type definitions created by their security professionals. Those tools are integrated with threats detection engines and ensure immediate data leakage prevention after their installation. Pre-defined set of sensitive data type definitions contain ready-for-use rules for PII (Personal Identifiable Information), intellectual property elements and other data types like credit cards, bank accounts, national identification numbers, social security number, etc. Users can either select which data type definitions from this set to apply on their endpoints or customize some of them for their own conditions, regulations or needs.

Flexible security policies wizards allow them also to define objects for content scanning including endpoints and groups, email senders and recipients, file and device types. In these tools events that trigger them can be selected - content copying to a removable storage device, uploading content in browsers and IM clients, sending data via email. In such events DLP can log the event, warn the user, block the transaction, quarantine or encrypt the content before sending.

Some software tools try to analyze and assess former communications from historical archives in protected system and to generate additional security rules from that gather knowledge that, with such learning approach, counteract more reliably to sensitive data leakage.

3 Applicability of Technologies for eGovernment Users Information Security Protection

There are some basic requirements which must meet in phase of development and implementation of security policies for identified or universal users of eGovernment services [12]:

- minimum need computer skills in eGovermnent users, which mainly will create and implement security policies with their own knowledge. The technologies they will use must be as transparent as is possible;
- minimum need of financial resources for implementing the technologies which will make them available for more citizens who must become eGovernment services users;
- possibilities for easy advance of implemented technologies with users experience and security needs evolution.

Based on information from international Endpoint Protection Platform vendors [4], [5], [7] and from surveys for endpoint security state [3], [6], [11] some conclusions about applicability of anti-theft and data leakage prevention technologies for protection of information security of eGovernment users can be done:

- To be applicable for eGovernment users, Ownership management services must be integrated in the eGovernment services themselves;
- To be more useful for eGovernment users, internal timer protection tool must be implemented as a software module, which is automatically installed when the users register themself for the respective class of eGovernment services. At this moment all protected user sensitive data must be automatically identified and the destructive routine must be transparently configured;
- It would be an advantage if the users can automatically unlock their computer system in case it was found after deactivation and if the user can authenticate himself with eSignature, by example;
- At this moment technologies for sensitive data leakage prevention are too complex for common eGovernment users and can be implemented only with gentle approach for DLP in protection of their information security. Tools who offer customizable set of pre-defined sensitive data type definitions are most suitable and perspective decisions in this direction;
- Some of advantages of DLP technologies can be reached only in integration with automatic and transparent identification of sensitive user data in the process of user registration for eGovernment services and their future protection. Security policy configuration wizards can give the possibilities for selection and customization of embedded sensitive data type definitions to apply, for event and action settings. They can automatically put in protection data such as credit and debit card numbers, bank accounts. national identification numbers, social insurance numbers, fiscal code number, postal adresses, phone numbers, passport details, email adresses, etc;
- Integration of DLP technologies with other services like e-mail, instant messaging, Skype, social networking, etc would be a challenge and a big advantage for eGovernment services users.

References

1. State of Internet security: protecting the network. Webroot software (2009)
2. Zeltser, L.: Emerging Internet security threats in 2009 (2009)

3. CSI Computer Crime and security survey. CSI (2009)
4. CISCO Midyear Security Report., CISCO (2009)
5. Symantec Endpoint Protection - value delivery research study. Symantec Corp. (2009)
6. Magic Quadrant for Endpoint protection platform. Gartner (2009)
7. Endpoint security and data protection. Sophos (2009)
8. Filkins, B., Radcliff, D.: Data leakage landscape: Where data leaks and how to apply next generation tools. SANS Institute (2008)
9. Mogull, R.: Is DLP keeping your data where it should be? Information Security magazine 2 (2009)
10. The evolution of endpoint security. Sophos (2009)
11. State of Endpoint. Ponemon institute (2009)
12. Palazov, A.: Policies and architectures for information security of citizens as users of the eGovernment. Sofia, Alternativi (2008)

Real-Time System for Assessing the Information Security of Computer Networks

Dimitrina Polimirova and Eugene Nickolov

National Laboratory of Computer Virology - Bulgarian Academy of Sciences,
1113 Sofia, Bulgaria, "Acad. Georgi Bonchev" Str., Block 8, Office 104
{polimira,eugene}@nlcv.bas.bg

Abstract. The report examines the possibility of establishing of real-time system for analysis and assessment of information security of computers, systems and networks in Internet/Intranet/Extranet environment, using TCP/IP protocols. In the paper are presented known information attacks. Separate classes of malicious software investigations are considered concerning different work platforms (produced by different Computing Systems), work environments (produced by different Browser Systems) and work places (produced by different Antimalware Systems). Methods that can be used to implement the systems are suggested. The capabilities of real-time systems are commented at the end of the paper.

Keywords: Data Security, Computer Security, Communication Security, Operating System Security, Web Security, Application Security.

1   The Problem

The modern information society requires the use of various types and configuration computers, systems and networks in TCP/IP environment. These computers, systems and networks are subject to permanent attacks with respect to their information security, which determines the need of investigation on methods and means for their protection.

A common strategy for protecting computers, systems and networks includes using of antivirus and security software. In addition the development of suitable for those computers, systems and networks security policy could be included.

Within the above research opportunities may apprise that it is appropriate to examine this problems of individual stages.

For the purpose of this paper can say that it is necessary to analyze the benefit of building of real-time system for assessment of information security of various types computers, systems and networks, exposed to one or more attacks, taking into account the impact of the protection methods in the face of various antivirus and security software.

Since the 70-th of century the problem for security and protection of computers, systems and networks has drawn developers' and constructors' attention in the area of information technology [1]. With the first malattack in the 60th of last century [2], a progress in the area of information security of computers,

J. Camenisch, V. Kisimov, and M. Dubovitskaya (Eds.): iNetSec 2010, LNCS 6555, pp. 123–133, 2011.

systems and networks is observed and requirements for their information security are increased. As a result for short time ideas significantly reducing the risk in the management of computers, systems and networks are realized.

2   Actuality

The main open problem, which can be placed within this paper, is related with the investigation of public known information attacks on one hand and most popular computing, browsers and antimalware *systems* on the other hand.

The main hypothesis will be linked with the ability to analyze and evaluate the effectiveness of a real-time system for assessing information security used as a means of determining the security policy with the lowest risk for individual computers, systems and networks.

Further analyses and investigations can be made towards precise planning of the economic costs of conducting a security policy for different configurations computers, systems and networks.

3   Goal and Tasks of the Investigation

The studies, which may be planned, should be linked with an analysis of the current state and development prospects of known information attacks, computing systems, browser systems and antimalware systems. Their scientific generalization in the form of real-time systems for assessing of information security is necessary because only in their mutual relations can be achieved the best analysis and therefore the best solutions for information security of computers, systems and networks.

3.1   The Main Goal

As a result of mentioned above the *main goal* of the paper can be specified: to analyze the effectiveness of the integration of the separate classes of malicious software investigations concerning different experimental work platforms, experimental work environments and experimental work places.

The integration is represented by a real-time system for presenting the obtained results with the help of which analysis and assessment of information security of computer systems and networks in the TCP/IP environment can be made.

3.2   Main Tasks

The following *main tasks* are set in reaching the goal:

1. To identify and systematize information attacks known until the moment of investigation;
2. To identify and systematize investigated:
   2.1 Operating Systems (OS), used by the Computing Systems;
   2.2 Browsers, used by the Browser Systems;
   2.3 Antivirus and security software, used by the Antimalware Systems.

4   Work Definitions

For the goal of this paper the following *work definitions* are proposed [3]:

1. As *information security* we will note the protection of the information in computers, systems and networks from a random or purposeful access of their resources aimed at reading, transferring (coping), modifying or destroying the information;
2. As *information attack* we will note any attempt to break the integrity of information objects that can be bit, byte, sector, file, directory system, browser systems, antimalware systems, operating systems, network systems, etc.;
3. As *computing system* we will note combination of hardware and software tools for solving specific application problems using preliminary prepared algorithmic solutions;
4. As *browser system* we will note software tools (programming tool), which is designed to perform requests for serving in client–server architecture and web functionality;
5. As *antimalware systems* we will note combination of different types of protecting system's components, which includes the following: anti-virus, anti-adware, anti-spyware, anti-trojan horse, anti-downloader, etc.

5   Analysis of Information Attacks

The information attacks can be divided into two main categories: malware and malattacks (Fig. 1). The main difference between them lies in the fact that in case of malware the direct participation of a user at the moment of the attack is missing, while in case of malattack the user's presence is required [4], [5].

Fig. 1. Main categories information attacks

The following attacks can be collected from the current information base of National Laboratory of Computer Virology of Bulgarian Academy of Sciences [6]. It collects information for the information attacks, which were carried out to a separate personal and/or corporate computers, and/or networks, and/or systems for the 2009. This is a generalization of the attacks, implemented in Bulgaria, Balkan Peninsula and south-east Europe.

5.1   Malware

In the category *MALWARE* are included 64 different information attacks, divided into 20 groups (Table 1.).

Table 1. Description of information attacks and their corresponding groups for the *MALWARE* category

Malware Groups	Single Information Attacks
I. Ads	(1) AdServer; (2) Adware; (3) Anarchie; (4) Banner; (5) Square news; (6) Investitial; (7) Superstitial; (8) Spam;
II. Browsers	(9) ActiveX; (10) BHO ; (11) Cookie; (12) Prefix of URL; (13) Related Info; (14) Scumware; (15) Ticker;
III. Metadata	(16) ADS (Alternate Data Streams); (17) Binder; (18) Downloader (Trojan Downloader); (19) Dropper;
IV. Joke	(20) Annoyance; (21) Joke;
V. Chat	(22) AOL Attack (America On Line Attack);
VI. Criminal Investigations	(23) Carnivore (DCS1000);
VII. Cracking	(24) Cracking; (25) Password Cracker;
VIII. Spying	(26) Spyware; (27) GUID; (28) IRC Bots; (29) Phishing; (30) Error Reporting Tool; (31) Smart Links; (32) Sniffing; (33) Toolbar; (34) nPnP; (35) WebBug; (36) GSM Pointer; (37) WAP Access Link;
IX. DoS, DDoS	(38) Flooder; (39) Mail Bomber; (40) Nuker; (41) Spoofing; (42) Bacterium;
X. Exploits	(43) Exploit;
XI. Hoaxes	(44) Hoax;
XII. Pop-Ups	(45) Pop-Over; (46) Pop-Under; (47) Pop-Up; (48) Pop-Roll; (49) Pop-Slider;
XIII. Scanners	(50) Port Scanner (IP Scanner); (51) Probe Tools; (52) RAT (Remote Administration Tool); (53) Riskware;
XIV. Keyboard modifiers	(54) Ansi Bomb; (55) Keylogger; (56) Mouselogger; (57) Screenlogger;
XV. Card Fishing	(58) Carding malware;
XVI. Dialer	(58) Dialer;
XVII. Computer Trojan Horses	(60) Computer Trojan Horses;
XVIII. Computer Backdoors	(61) Computer Backdoors;
XIX. Computer Worms	(62) Computer Worms;

5.2   Malattacks

In the category *MALATTACKS* are included 24 different information attacks, divided into 13 groups (Table 2.).

Table 2. Description of information attacks and their corresponding groups for the *MALATTACK* category

Malattacks Groups	Single Attacks
XXI. Using accessible information	(65) Audit Trail; (66) Traffic Analysis;
XXII. Overflow	(67) Buffer Overflow;
XXIII. Vulnerabilities	(68) CGI (Common Gateway Interface) Vulnerabilities; (69) Hijacking; (70) Packet Attacks;
XXIV. Content	(71) Content Attacks;
XXV. Data Encapsulation	(72) Data Driven;
XXVI. Denial of Service	(73) DDoS (Distributed Denial of Service); (74) Flooding;
XXVII. Spoofing	(75) DNS Spoofing; (76) EFT Spoofing; (77) Ethernet Spoofing; (78) IP Spoofing; (79) Screen Spoofing; (80) SET Spoofing; (81) TCP Dump Spoofing; (82) Trace Route Spoofing; (83) Tunnel Spoofing;
XXVIII. CrackPasswd	(84) HTCrackPasswd;
XXIX. Wire/Wireless Phones	(85) Phreaking;
XXX. Physical Analyze Devices	(86) Physical Perimeter Penetration;
XXXI. Social Engineering	(87) Social Engineering;
XXXII. EMI/RFI Intercepts	(88) Wireless Intercepts;
XXXIII. Zombie Computers	(89) Zombies;

Note: Roman numbers in are used later as identifiers of the names of the information attacks groups.

6   Work Platforms, Environments and Places

6.1   Work Platforms

Different experimental work platforms can be investigated for experimental study and analysis of malware for computing systems. The most popular currently using computing systems are based on the operating systems Windows, Linux and Mac. Therefore, for the purposes of this paper, they can be described as basic. On this basis, a reasonable mix of investigations into each one of them may seek and along with it to seek a reasonable summary of their mutual influence.

Fig. 2 shows a percentage distribution of accomplished attacks to Windows OS, Linux OS and Mac OS.

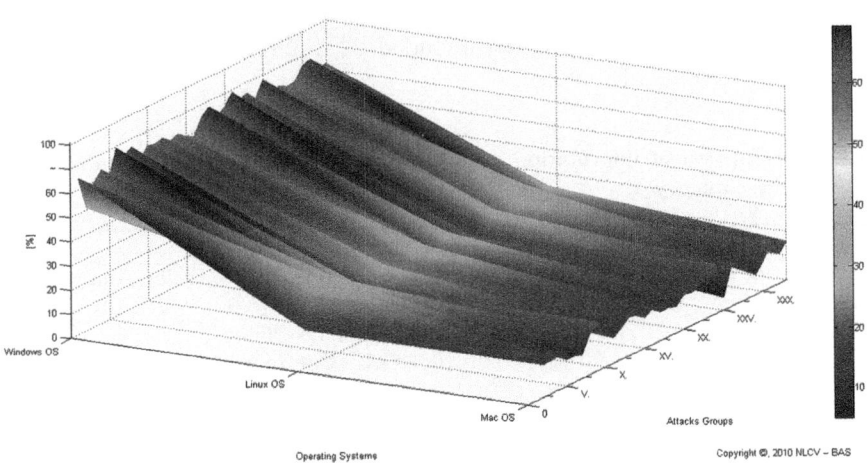

Fig. 2. Percentage distribution of accomplished attacks to Windows, Linux, and Mac

6.2   Experimental Work Places

With respect to the work environments for investigating and analyzing malware for Browser Systems may say that at the current state of information threats can be assessed that the main problem for the security of home, corporate, government networks is the type of the installed browser. Achievements in this area of the various developers are measured not by days, but by hours. With regard to this is sensibly to analyze some chosen from the top 10 classification browsers for the different operating systems.

Fig. 3 shows chosen top 10 the most popular browsers for Bulgaria, Balkan Peninsula and south-east Europe for the main operating systems.

Fig. 4, Fig. 5, and Fig. 6 show percentage distribution of information attacks, accomplished to the separate operating systems with respect to the separate Browser Systems.

6.3   Work Places

With respect to the work places for investigating and analyzing malware based on Antimalware Systems may say that when assessing the information security of the endpoints in the information structure is necessary to report the presence/absent of specialized antivirus and security tools. Depending on the type and nature of the activity and depending on the available funds, different antivirus and security solutions with respect to their functionality and with respect to the end user competent can be chosen. Therefore it is extremely difficult to find a widely applicable antivirus and security solution. This necessitates creating of top 10 antivirus and security solutions classifications which can be used for different investigations.

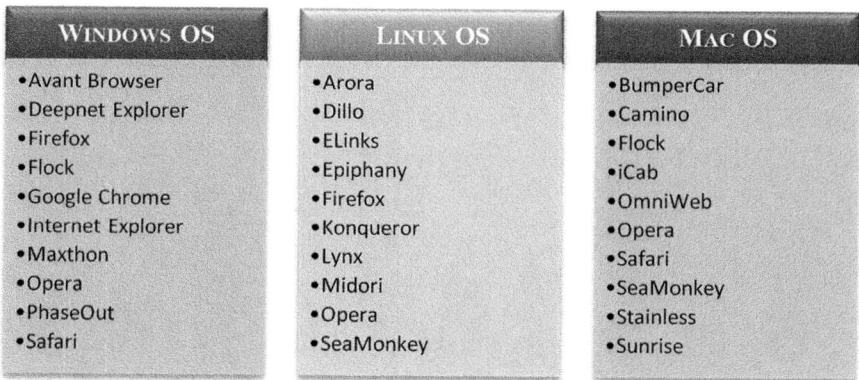

Fig. 3. Top 10 most popular browsers for Windows, Linux and Mac

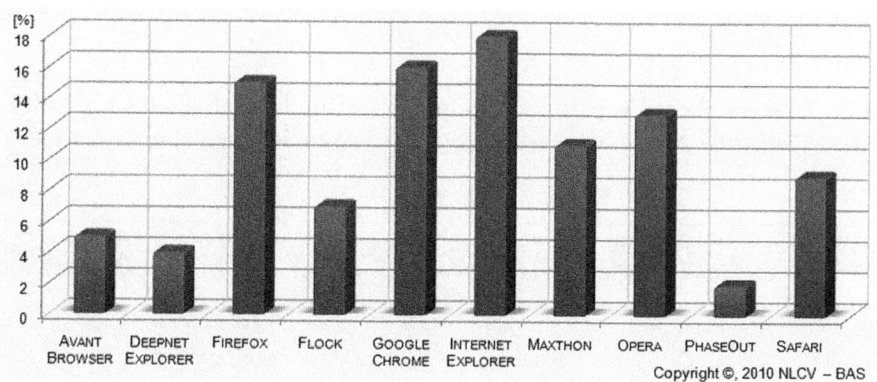

Fig. 4. Percentage distribution of accomplished attacks to Windows OS with respect to the separate Browser Systems

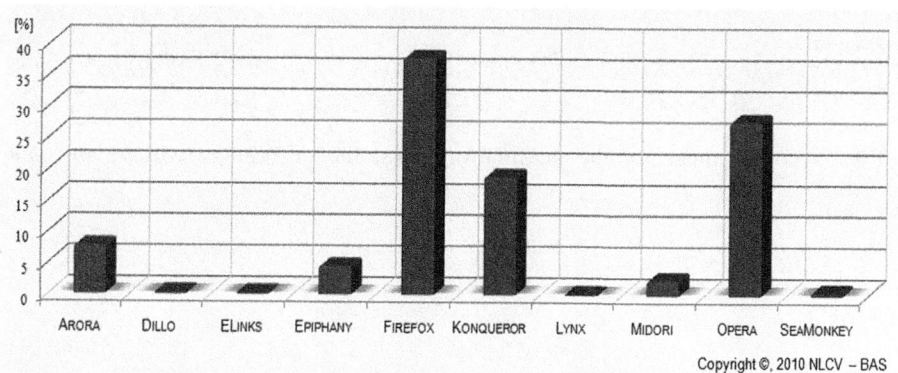

Fig. 5. Percent distribution of accomplished attacks to Linux OS with respect to the separate Browser Systems

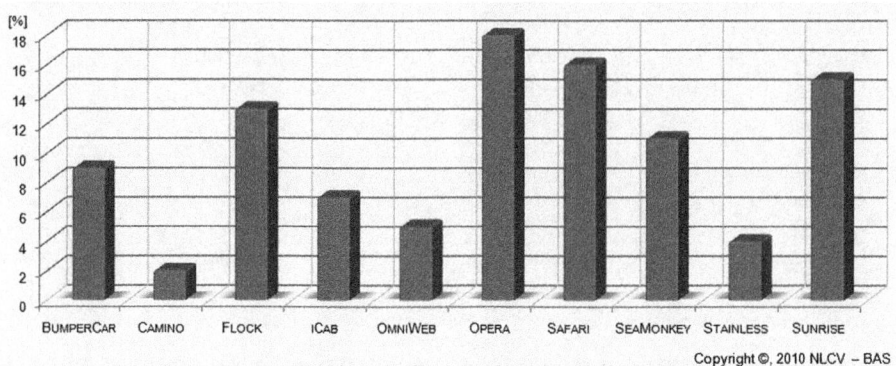

Fig. 6. Percent distribution of accomplished attacks to Mac OS with respect to the separate Browser Systems

Fig. 7 shows chosen top 10 most popular Antimalware Systems for Bulgaria, Balkan Peninsula and south-east Europe for the main operating systems.

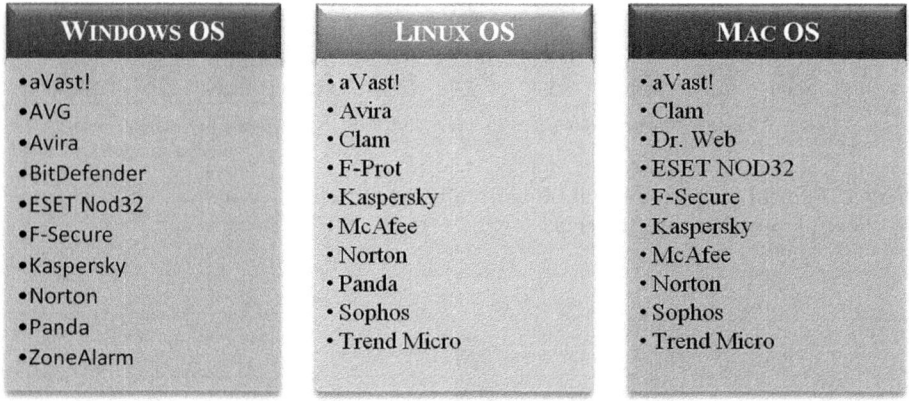

Fig. 7. Top 10 most popular Antimalware Systems for Windows, Linux and Mac

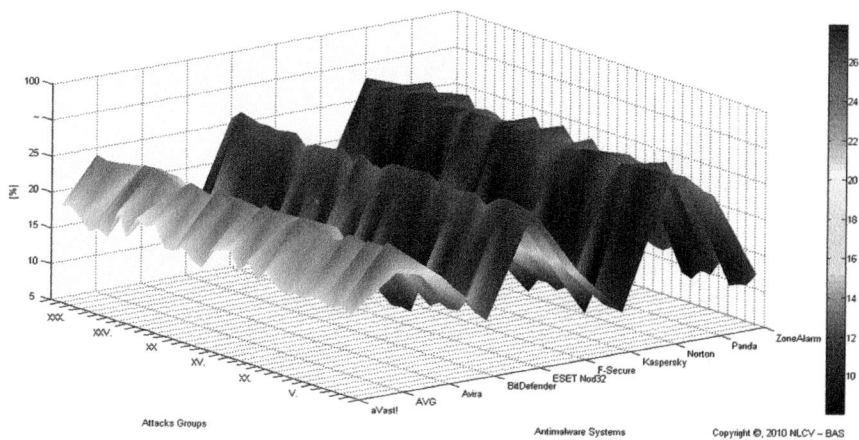

Fig. 8. Percentage distribution of successful accomplished attacks groups to the Windows OS with respect to the separate Antimalware Systems

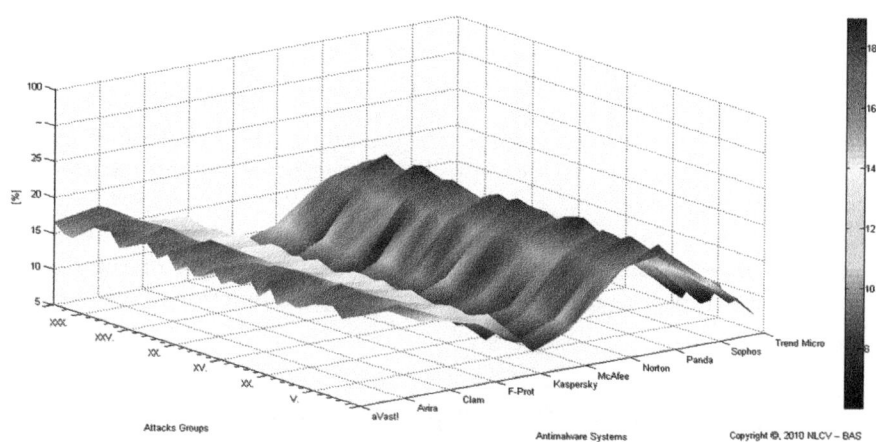

Fig. 9. Percentage distribution of successful accomplished attacks groups to the Linux OS with respect to the separate Antimalware Systems

Fig. 8, Fig. 9, Fig. 10 show the percentage distribution of successful accomplished attacks groups to the separate operating systems with respect to the separate Antimalware Systems.

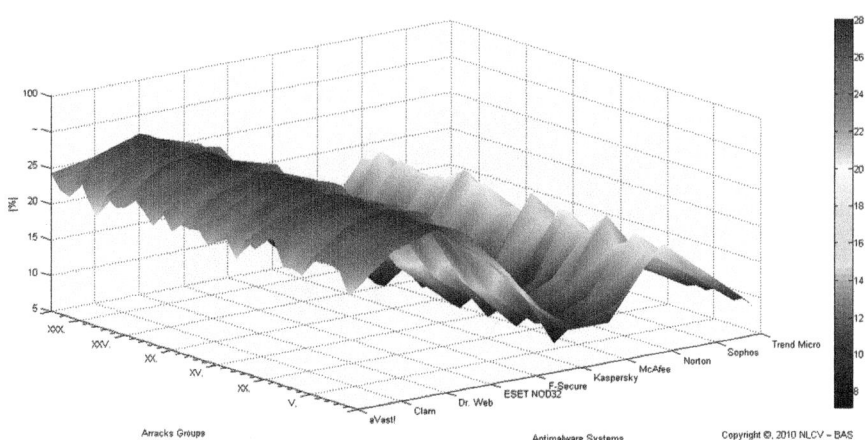

Fig. 10. Percentage distribution of successful accomplished attacks groups to the Mac OS with respect to the separate Antimalware Systems

7   Used Methods

Methods which can be used for the realization of real-time system for assessment of information security of computer networks can include:

1. Creating a reference *binary sequences* describing the behavior of malware through which planned actions in different platforms, environments and locations are carried out;
2. Creating a *data containers* for cyclic accumulation, processing and archiving;
3. Creating a *graphical environment* for online visualization of the obtained results with respect to the selected parameters as a function of other selected parameters.

8   Possibilities of the Real-Time System

The real-time system for assessing the information security will be able to provide a possibility for:

1. Assessing the *velocity propagation* of malware, the *kind* of attacked objects and the *methods* for reducing the impact of malware over TCP/IP environment.

    The methods for reducing the impact include:

    1.1 detection and real-time protection;
    2.2 detection and real-time cleaning;
    2.3 detection and real-time immunization;
    2.4 detection and real-time quarantine.

2. *Assessing the sustainability* of the impact of different malware on various types of information objects in different platforms, environments and locations;
3. *Assessing the applicability and approbation* of different types and kinds of commercial antivirus and security solutions.

    For each triple relation *platform–environment–place*, the assessment for applicability gives:

    3.1 the most security solution (with the lowest risk);

    3.2 the most economical solution (at acceptable risk).

The assessment for approbation gives an opportunity for each selected place, which claimes a certain volume functionality applicable to a certain combination *platform–environment*, to be officially confirmed, that it has the announced functionality.

9 Conclusions and Recommendations

The chosen formulation for investigations of malware for different computing systems (presented by operating systems), browser systems (presented by the top 10 most popular browsers) and antimalware systems (presented by the top 10 most popular antivirus and security software) contains the necessary potential for extensive research in this and other neighboring areas.

The results obtained by the real-time system for assessing the information security, give an opportunity for concrete planning of security policy of different configurations computers, systems and networks.

Conditions for precise planning of economic expenses, related to the performing of security policy for determinate configuration computer, system and networks, are created.

References

1. Denning, D.E.: A lattice model of secure information flow. Communications of the ACM 19(5), 236–243 (1976)
2. The St. Petersburg Times,
   http://www.sptimes.com/Hackers/history.hacking.html
3. Brotby, W.K.: Information Security Management Metrics: A Definitive Guide to Effective Security Monitoring and Measurement, pp. 7–8. CRC Press, Boca Raton (2009)
4. Parsons, J.J., Oja, D.: New Perspectives Computer Concepts 2010: Introductory. Cengage Learning, p. 162 (2009)
5. Radhamani, G., Rao, R.: Web Services Security and E-business, p. 115, p. 25, Global (2007)
6. National Laboratory of Computer Virology – BAS, National Cybersecurity Portal,
   http://ncs.nlcv.bas.bg/

Evidential Notions of Defensibility and Admissibility with Property Preservation

Raphael C.-W. Phan, Ahmad R. Amran,
John N. Whitley, and David J. Parish

Loughborough University
within the High Speed Networks (HSN) Lab
of the Electronic and Electrical Engineering department
LE11 3TU, UK
{r.phan,a.r.amran,j.n.whitley,d.j.parish}@lboro.ac.uk

Abstract. For security-emphasizing fields that deal with evidential data acquisition, processing, communication, storage and presentation, for instance network forensics, border security and enforcement surveillance, ultimately the outcome is not the technical output but rather physical prosecutions in court (e.g. of hackers, terrorists, law offenders) or counter-attack measures against the malicious adversaries.

The aim of this paper is to motivate the research direction of formally linking these technical fields with the legal field. Notably, deriving technical representations of evidential data such that they are useful as evidences in court; while aiming that the legal parties understand the technical representations in better light. More precisely, we design the security notions of evidence processing and acquisition, guided by the evidential requirements from the legal perspective; and discuss example relations to forensics investigations.

1 Motivation

For the security fields that involve evidential data acquisition (e.g. monitoring or surveillance), processing, communication, storage and presentation (or reconstruction), such as network forensics [6,9], border security and enforcement surveillance, the ultimate outcome is not so much the technical output but rather the physical prosecutions in court (e.g. of hackers, terrorists, law offenders) or counter-attack measures [10] against the adversaries behind malicious attacks.

This paper motivates the research direction of formally linking the above-mentioned technical fields with the legal field. The approach we suggest is to design the relevant security notions and the technical methods of evidence processing and construction [7] guided by the evidential requirements from the legal perspective.

J. Camenisch, V. Kisimov, and M. Dubovitskaya (Eds.): iNetSec 2010, LNCS 6555, pp. 134–139, 2011.

2   Directions

In more detail, we advocate the need to treat the following particular research directions, all of which are open problems to date.

- Formal definitions [12] of evidential notions: we classify evidential usefulness in terms of how usable it is as the evidence (define this as *admissibility*) and whether the processes applied to the evidence from crime scene to court are legally appropriate (define this as *defensibility*). Doing so provides the framework to guide the design of technical evidential collection and construction, as well as gives a sound mapping between the technical and legal fields so to avoid ambiguity during the transition from one field to another and/or throughout the life cycle of the evidence. This is crucial as defensibility requirements dictate that continuity of evidence be ensured from crime scene (or observed event) through to court room.
- Design of property-preserving processes for evidential data e.g. acquisition, duplication, storage, reconstruction: it is important that the evidence have continuity [14] from crime scene to court. Therefore, as the evidence is processed along the way, there is a need to ensure that the processes preserve the evidentiary properties [8] of the evidence such as provenance [14], integrity and admissibility.
- Design of abstract evidential data constructs satisfying legal admissibility requirements.

In this paper, we treat the first research direction. The last two directions are our on-going research.

3   Notions

We first define the primitive security properties desired for evidential data. We then define our main evidential notions, namely defensibility (dealing with the validity of the *process*) and admissibility (dealing with the validity of the *final evidence* presented to court).

3.1   Primitive Properties

Integrity, (source) authentication and linkability are the underlying properties that form the base of wider notions including defensibility and admissibility.

Integrity and Source Authentication. The integrity property and source authentication property of some evidential data can be based on cryptographic notions of integrity e.g. INT [1], and unforgeabiilty (UF) [4].

Linkability. Let $w(E^S, ID)$ denote a function that evaluates the weight of evidence E^S at state S of the evidence life cycle with respect to how much it links to a particular individual of identity ID. An evidence E^S is said to be *linkable* (LNK) to a particular person of identity ID if the weight of the evidence $w(E^S, ID) > \varepsilon$ for some negligible ε.

3.2 Defensibility

This notion can be defined to include the preservation of the properties of integrity, data provenance (a.k.a. chain of custody) and relevance (in terms of necessity, and linkability to the suspect and to the crime or event).

Property Preservation. The gist of the notion of property preservation (PPr) relates to the state of the evidential data E still retaining a particular property P even after undergoing a function $f(\cdot)$.

Let $f(\cdot)$ denote a function $\in$ {acquisition, processing, storage, communication, presentation, reconstruction} operating on some evidence E^S at state S. See Fig. 1.

If E^S exhibits a property P, denoted as $\Pr[E^S \mapsto P] > \varepsilon$ for some negligible ε, then $f(\cdot)$ is said to be property-preserving in the sense of P if $f(E^S)$ also exhibits property P, i.e. $\Pr[f(E^S) \mapsto P] > \varepsilon$.

Define an adversarial game where adversary A interacts with a forensics challenger and is allowed oracle access to any $f(\cdot) \in$ {acquisition, processing, storage, communication, presentation, reconstruction}. Furthermore, A is given samples of some evidence E^S exhibiting property P. The game ends when A outputs an evidence $\tilde{E}^S$.

Define A's advantage in winning the game as

$$\mathsf{Adv}_A^{\mathsf{PPr}} = ((\Pr[E^S \mapsto P] > \varepsilon) \wedge (\Pr[f(E^S) \mapsto P] < \varepsilon)) \vee$$
$$((\Pr[E^S \mapsto P] < \varepsilon) \wedge (\Pr[f(E^S) \mapsto P] > \varepsilon)).$$

A function is property-preserving in the sense of P if $\mathsf{Adv}_A^{\mathsf{PPr}} < \varepsilon$.

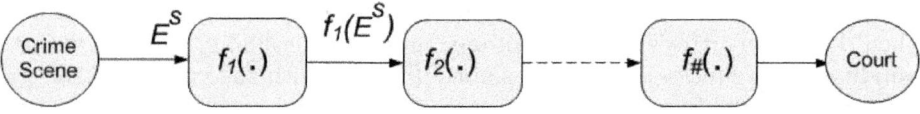

Fig. 1. Evidential data going through multiple processes from crime scene to court

The adversarial winning condition here captures the case where a function $f(\cdot)$ negates the existence of an evidence E's property P.

Provenance. This requires the integrity maintenance of custody information i.e. information about which party is responsible (holds the custody) over the evidence, at each state of the process from crime (or event) scene to court.

Relevance. For the relevance notion, we emphasize that besides capturing how the necessary evidential data are acquired and processed, as well as including linkability, the notion should also capture the assurance that non-necessary data is not represented e.g. private data that will intrude the privacy of non-involved passers-by who just happen to be there. This is an interesting open problem, since for instance current enforcement monitoring systems such as CCTVs and border security controls do not formally assure privacy preservation and this notion of relevance.

3.3 Admissibility

This notion emphasizes on the verification of the final evidence in terms of its integrity, provenance and relevance. Herein, adversarial games can be defined for each of the properties. While integrity will be more straightforward, the provenance property should include a measure of the evidential data origin including source (e.g. IP address for network forensics, or identification details of the suspect) and event/crime location details, whereas the relevance property is as discussed above in subsection 3.2.

4 Exemplification

To be more illustrative, we present some brief instances to exemplify the applicability of Section 3's notions in network forensics.

4.1 Integrity and Linkability

Consider an instance of telecommunications service in an organisation, telephone calls taking place on or through corporate switchboards (PABXs). Such switchboards routinely provide data about the numbers called and the time and duration of calls. They do so as a means to monitor costs for outgoing calls and to check on service quality with respect to internal calls. The logs produced can be of considerable value in many kinds of investigations. Some businesses routinely record phone calls as a check against disputed transactions, or to see whether their employees are misbehaving (misuse, corporate sabotage, terrorism etc).

In a forensic situation, the immediate and important issue is to be able to demonstrate that the logs and/or recordings are reliable and have not been tampered with. It is helpful to be able to say something about the specific PABX and what logging facilities exist. There should be some statements about how they were collected, by whom, what precautions were taken, and how selections of data were made. The data should also be subjected to some form of integrity check, as a guard against post-capture tampering and forgery. They can specifically be linked to alleged individuals and circumstances. Thus, any evidential data will be considered for weight of fact with respect to its persuasiveness and probative value.

4.2 Provenance, Weight of Evidence and Property Preservation

In any type of investigation, the forensics challenger must follow an investigation process. That process begins with the step of assessing the case and documenting them in an effort to identify the crime and the location of the evidence. An evidence's chain of custody (data provenance) must be prepared to know who handled the evidence, and every step taken by the forensic investigator must be documented for inclusion in the final report.

Sometimes a computer and its related evidence can determine the chain of events leading to a crime for the investigator as well as provide the evidence which can lead to conviction. For instance, in a network attack instance; each adversary could have his own unique motives, methods and skills. He begins with little or no information about the target. However, depending on his skills, he might be able to construct a detailed roadmap that may enable the adversary to compromise the target. The adversary's approach generally covers several of these seven steps (or all of them based on their motives, methods and skills): 1) Perform a footprint analysis e.g., IP address, domain name, location, subsidiaries etc; 2) Enumerate information e.g., type and version of OS, ftp, mail and services running etc; 3) Attack the network and penetrate systems that have high vulnerabilities e.g., knowledge gained from published Common Vulnerability and Exposure (CVEs); 4) Escalate privileges to have access control e.g., social engineering and/or brute force; 5) Raid information and user records; 6) Install backdoor to circumvent trusted programs; 7) Leverage the compromised system. Any evidential data of any of these or more would cause a certain degree of severity to the damage done, and this relates to the weight of evidence.

Finding the evidence, discovering relevant data, eradicating external avenues of alteration, gathering the evidence, and preparing a chain of custody are processes where evidential properties need to be preserved through to court. A forensics challenger must ensure that evidential data still exhibits a particular property (e.g. integrity) from when they are first gathered and seized at the crime scene, through the processes of the chain of custody preparation, evidence transportation to and/or storage at forensics labs, to ultimately the court room.

5   Remarks

Defensibility methods could be designed as essentially protocols that have underlying security mechanisms for integrity checking (e.g. message authentication codes or hashes keyed by secrets), source authentication (e.g. digital signatures) and traceability. Parties are used to model the interaction from one intermediary spatial point or representative form to another during the course of the evidence from crime scene through to court.

Admissible methods could be designed as mechanisms rather than protocols as they do not involve interaction among parties but rather apply to the evidential properties of the final evidence.

Having explicit and non-ambiguous notions and processes allow the proper mapping from technical evidence processing of monitored scene through to legal courts and enforcement, and this will lead to effective evidential systems where the processing of evidential data is non-wasteful and fit for purpose.

References

1. Bellare, M., Namprempre, C.: Authenticated encryption: Relations among notions and analysis of the generic composition paradigm. In: Okamoto, T. (ed.) ASI-ACRYPT 2000. LNCS, vol. 1976, pp. 531–545. Springer, Heidelberg (2000)
2. Caloyannides, M.A.: Forensics is so 'Yesterday'. IEEE Security and Privacy 7(2), 18–25 (2009)
3. Carrier, B.D.: Digital Forensics Works. IEEE Security and Privacy 7(2), 26–29 (2009)
4. Goldwasser, S., Micali, S., Rivest, R.: A Digital Signature Scheme Secure against Adaptive Chosen-Message Attacks. SIAM Journal on Computing 17(2), 281–308 (1988)
5. Kearsley, A.J.: Electronic Document Management: Legal Admissibility of Evidence Held in Digital Form. Computer Law & Security Report 15(3), 185–187 (1999)
6. Kenneally, E.E.: Digital Logs - Proof Matters. Digital Investigation 1, 94–101 (2004)
7. Kenneally, E.E., Brown, C.L.T.: Risk Sensitive Digital Evidence Collection. Digital Investigation 2, 101–119 (2005)
8. Mocas, S.: Building Theoretical Underpinnings for Digital Forensics Research. Digital Investigation 1, 61–68 (2004)
9. Nikkel, B.J.: Improving Evidence Acquisition from Live Network Sources. Digital Investigation 3, 89–96 (2006)
10. Phan, R.C.-W., Whitley, J.N., Parish, D.J.: Adversarial Security: Getting to the Root of the Problem. In: Camenisch, J., Kisimov, V., Dubovitskaya, M. (eds.) iNetSec 2010. LNCS, vol. 6555, pp. 47–55. Springer, Heidelberg (2010)
11. Reed, C.: The Admissibility and Authentication of Computer Evidence - a Confusion of Issues. Computer Law & Security Report 6(2), 13–16 (1990)
12. Rogaway, P.: Practice-Oriented Provable Security and the Social Construction of Cryptography. In: Eurocrypt 2009, invited talk (May 6, 2009)
13. Solon, M., Harper, P.: Preparing Evidence for Court. Digital Investigation 1, 279–283 (2004)
14. Turner, P.: Digital Provenance - Interpretation, Verification and Corroboration. Digital Investigation 2, 45–49 (2005)

Cloud Infrastructure Security

Dimiter Velev[1] and Plamena Zlateva[2]

[1] University of National and World Economy,
UNSS - Studentski grad, 1700 Sofia, Bulgaria
dvelev@unwe.acad.bg
[2] Institute of Control and System Research - Bulgarian Academy of Sciences
Acad. G. Bonchev Str., Bl. 2, P.O. Box 79, 1113 Sofia, Bulgaria
plamzlateva@abv.bg

Abstract. Cloud computing can help companies accomplish more by eliminating the physical bonds between an IT infrastructure and its users. Users can purchase services from a cloud environment that could allow them to save money and focus on their core business. At the same time certain concerns have emerged as potential barriers to rapid adoption of cloud services such as security, privacy and reliability. Usually the information security professionals define the security rules, guidelines and best practices of the IT infrastructure of a given organization at the network, host and application levels. The current paper discusses miscellaneous problems of providing the infrastructure security. The different aspects of data security are given a special attention, especially data and its security. The main components of cloud infrastructure security are defined and the corresponding issues and recommendations are given.

Keywords: cloud, service, infrastructure, IaaS, security, data.

1   Introduction to Cloud Computing Basics

Currently one of the major topics of many information technology discussions is cloud computing and the key point in them is cloud computing security. Usually conversations focus on all standard security advantages, disadvantages and requirements. Nevertheless the fact the most common security measures protect data from loss, unauthorized access, integrity disruption, etc., there are other necessary and important characteristics of any IT infrastructure that must implemented in a much more serious way. One of those structures is the cloud infrastructure.

1.1   Cloud Computing Definition

Cloud computing is an on-demand service model for IT provision based on virtualization and distributed computing technologies [1], [9], [10]. Typical cloud computing providers deliver common business applications online as services which are accessed from another web service or software like a web browser,

J. Camenisch, V. Kisimov, and M. Dubovitskaya (Eds.): iNetSec 2010, LNCS 6555, pp. 140–148, 2011.
© IFIP International Federation for Information Processing 2011

while the software and data are stored on servers. The abstraction of computing, network and storage infrastructure is the foundation of cloud computing. The infrastructure is a service, and its components must be readily accessible and available to the immediate needs of the application stacks it supports. Cloud computing removes the traditional application silos within the data center and introduces a new level of flexibility and scalability to the IT organization. This flexibility helps address challenges facing enterprises and IT service providers that include rapidly changing IT landscapes, cost reduction pressures, and focus on time to market. Cloud users are maybe identified as follows [1]:

- Individual consumers;
- Individual businesses;
- Start-ups;
- Small and medium-size businesses;
- Enterprise businesses

Cloud computing architectures offer to its users numerous advantages that can be briefly summarized to [2]:

- reduced cost since services are provided on demand with pay-as-you-use billing system;
- highly abstracted resources;
- instant scalability and flexibility;
- instantaneous provisioning;
- shared resources, such as hardware, database, etc.;
- programmatic management through API of Web services;
- increased mobility - information is accessed from any location.

1.2   Cloud Computing Categories

The following cloud computing categories have been identified and defined in the process of cloud development [1], [10]:

- Infrastructure as Service (IaaS): provides virtual machines and other abstracted hardware and operating systems which may be controlled through a service Application Programming Interface (API). IaaS includes the entire infrastructure resource stack from the facilities to the hardware platforms that reside in them. It incorporates the capability to abstract resources as well as deliver physical and logical connectivity to those resources. IaaS provides a set of APIs which allow management and other forms of interaction with the infrastructure by consumers.
- Platform as a Service (PaaS): allows customers to develop new applications using APIs, implemented and operated remotely. The platforms offered include development tools, configuration management and deployment platforms. PaaS is positioned over IaaS and adds an additional layer of integration with application development frameworks and functions such as database, messaging, and queuing that allow developers to build applications for the platform with programming languages and tools are supported by the stack.

- Software as a Service (SaaS): is software offered by a third party provider, available on demand, usually through a Web browser, operating in a remote manner. Examples include online word processing and spreadsheet tools, CRM services and Web content delivery services. SaaS in turn is built upon the underlying IaaS and PaaS stacks and provides a self-contained operating environment used to deliver the entire user experience including the content, its presentation, the applications and management capabilities.
- Multi-Tenancy: the need for policy-driven enforcement, segmentation, isolation, governance, service levels and billing models for different consumer constituencies. Consumers might utilize a public cloud provider's service offerings or actually be from the same organization, but would still share infrastructure.

1.3   Cloud Deployment Models

The cloud services can be implemented in four deployment models [1], [10]:

- Public Cloud. The cloud infrastructure is made available to the general public or large industry group and is owned by an organization selling cloud services.
- Private Cloud. The cloud infrastructure is operated entirely for a single organization. It may be managed by the organization or a third party, and may exist on-premises or off-premises.
- Community Cloud. The cloud infrastructure is shared by several organizations and supports a specific community. It may be managed by the organizations or a third party, and may exist on-premises or off-premises.
- Hybrid Cloud. The cloud infrastructure is a composition of two or more clouds (private, community or public) that are bound together by standardized or proprietary technology that enables portability of data and application.

2   Risks and Security Concerns with Cloud Computing

Many of the cloud computing associated risks are not new and can be found in the computing environments. There are many companies and organizations that outsource significant parts of their business due to the globalization. It means not only using the services and technology of the cloud provider, but many questions dealing with the way the provider runs his security policy. After performing an analysis the top threats to cloud computing can be summarized as follows [3], [7]:

- Abuse and Unallowed Use of Cloud Computing;
- Insecure Application Programming Interfaces;
- Malicious Insiders;
- Shared Technology Vulnerabilities;
- Data Loss and Leakage
- Account, Service and Traffic Hijacking;
- Unknown Risk Profile.

It has been established that the most common topics related with cloud computing risk at present include [6], [11]:

- The cloud provider takes responsibility for information handling which is a critical part of the business. Failure to perform to agreed service levels can impact not only confidentiality but also availability.
- The dynamic nature of cloud computing may result in confusion as to where information actually resides. This may create delays when information retrieval is required.
- Third-party access to sensitive information creates a risk of compromise to confidential information. This can pose a significant threat to ensuring the protection of intellectual property and trade secrets.
- Public clouds allow high-availability systems to be developed at service levels often impossible to create in private networks. Compliance to regulations and laws in different geographic regions can be a challenge for business.
- Due to the dynamic nature of the cloud, information may not be located in the event of a disaster immediately. Business continuity and disaster recovery plans must be well documented and tested. Recovery time objectives should be stated in the contract. When faced with the paradigm change and nature of services provided through cloud computing, there are many challenges for cloud providers [6]. Some of the major security issues that will need to be addressed are [4], [5], [12]:
- Transparency - Service providers must provide for the existence of effective and robust security controls, assuring customers that their information is properly secured against unauthorized access, change and
- Privacy - With privacy concerns growing across the globe it will be imperative for cloud computing service providers to prove to existing and prospective customers that privacy controls are in place and demonstrate their ability to prevent, detect and react to breaches in a proper manner. Information and reporting communication lines need to be organized and agreed before service provisioning starts. These communication channels should be tested periodically during operations.
- Compliance - Most organizations must comply with a wide set of laws, regulations and standards. There are concerns with cloud computing that data may not be stored in one place and may not be easily retrievable. Audits completed by legal, standard and regulatory authorities demonstrate that there can be plenty of problems. When using cloud services there is no guarantee that a certain company can get its information when needed, and even some providers are reserving the right to withhold information from authorities.
- Transborder information flow - When information can be stored anywhere in the cloud, the physical location of the information can become an issue. Physical location rules jurisdiction and legal obligation. Country laws governing personally identifiable information may vary significantly.
- Certification - Cloud computing service providers will need to provide their customers assurance that they operate surely. Independent assurance from third-party audits and/or service auditor reports should be a vital part of any assurance program.

3 Cloud Security Principles

Public cloud computing requires a security model that coordinates scalability and multi-tenancy with the requirement for trust. As enterprises move their computing environments with their identities, information and infrastructure to the cloud, they must be willing to give up some level of control. In order to do so they must be able to trust cloud systems and providers, as well as to verify cloud processes and events. Important building blocks of trust and verification relationships include access control, data security, compliance and event management - all security elements well understood by IT departments today, implemented with existing products and technologies, and extendable into the cloud. The cloud security principles comprise three categories: identity, information and infrastructure.

3.1 Identity Security

End-to-end identity management, third-party authentication services and identity must become a key element of cloud security. Identity security keeps the integrity and confidentiality of data and applications while making access readily available to appropriate users. Support for these identity management capabilities for both users and infrastructure components will be a major requirement for cloud computing and identity will have to be managed in ways that build trust. It will require:

- Stronger authentication: Cloud computing must move beyond authentication of username and password, which means adopting methods and technologies that are IT standard IT such as strong authentication, coordination within and between enterprises, and risk-based authentication, measuring behavior history, current context and other factors to assess the risk level of a user request.
- Stronger authorization: Authorization can be stronger within an enterprise or a private cloud, but in order to handle sensitive data and compliance requirements, public clouds will need stronger authorization capabilities that can be constant throughout the lifecycle of the cloud infrastructure and the data.

3.2 Information Security

In the traditional data center, controls on physical access, access to hardware and software and identity controls all combine to protect the data. In the cloud, that protective barrier that secures infrastructure is diffused. The data needs its own security and will require [5], [14]:

- Data isolation: In multi-tenancy environment data must be held securely in order to protect it when multiple customers use shared resources. Virtualization, encryption and access control will be workhorses for enabling varying degrees of separation between corporations, communities of interest and users.

- Stronger data security: In existing data center environments the role-based access control at the level of user groups is acceptable in most cases since the information remains within the control of the enterprise. However, sensitive data will require security at the file, field or block level to meet the demands of assurance and compliance for information in the cloud.
- Effective data classification: Enterprises will need to know what type of data is important and where it is located as prerequisites to making performance cost-benefit decisions, as well as ensuring focus on the most critical areas for data loss prevention procedures.
- Information rights management: it is often treated as a component of identity on which users have access to. The stronger data-centric security requires policies and control mechanisms on the storage and use of information to be associated directly with the information itself.
- Governance and compliance: A major requirement of corporate information governance and compliance is the creation of management and validation information - monitoring and auditing the security state of the information with logging capabilities. The cloud computing infrastructures must be able to verify that data is being managed per the applicable local and international regulations with appropriate controls, log collection and reporting.

3.3 Security Compromises between the Three Cloud Deployment Models

The following security compromises between the three cloud deployment models have been identified [7], [10]:

- SaaS provides the most integrated functionality built directly into the offering, with the least consumer extensibility, and a relatively high level of integrated security since at the least the provider bears a responsibility for the security.
- PaaS is intended to enable developers to build their own applications on top of the platform. As a result it tends to be more extensible than SaaS, at the expense of customer ready features. This tradeoff extends to security features and capabilities, where the built-in capabilities are less complete, but there is more flexibility to layer on additional security.
- IaaS provides few if any application-like features, but enormous extensibility. This generally means less integrated security capabilities and functionality beyond protecting the infrastructure itself. This model requires that operating systems, applications, and content be managed and secured by the cloud consumer.

4 Infrastructure Security

IaaS application providers treat the applications within the customer virtual instance as a black box and therefore are completely indifferent to the operations and management of a applications of the customer [13]. The entire pack (customer application and run time application) is run on the customers' server on provider infrastructure and is managed by customers themselves. For this reason it is important to note that the customer must take full responsibility for securing their cloud deployed applications [7], [8], [12].

- Cloud deployed applications must be designed for the internet threat model.
- They must be designed with standard security countermeasures to guard against the common web vulnerabilities.
- Customers are responsible for keeping their applications up to date - and must therefore ensure they have a patch strategy to ensure their applications are screened from malware and hackers scanning for vulnerabilities to gain unauthorized access to their data within the cloud.
- Customers should not be tempted to use custom implementations of Authentication, Authorization and Accounting as these can become weak if not properly implemented.

The foundational infrastructure for a cloud must be inherently secure whether it is a private or public cloud or whether the service is SAAS, PAAS or IAAS. It will require [7], [9]:

- Inherent component-level security: The cloud needs to be architected to be secure, built with inherently secure components, deployed and provisioned securely with strong interfaces to other components and supported securely, with vulnerability-assessment and change-management processes that produce management information and service-level assurances that build trust.
- Stronger interface security: The points in the system where interaction takes place (user-to-network, server-to application) require stronger security policies and controls that ensure consistency and accountability.
- Resource lifecycle management: The economics of cloud computing are based on multi-tenancy and the sharing of resources. As the needs of the customers and requirements will change, a service provider must provision and decommission correspondingly those resources - bandwidth, servers, storage and security. This lifecycle process must be managed in order to build trust.

The infrastructure security can be viewed, assessed and implemented according its building levels - the network, host and application levels [7], [11].

4.1   Infrastructure Security – The Network Level

When looking at the network level of infrastructure security, it is important to distinguish between public clouds and private clouds. important to distinguish between public clouds and private clouds. With private clouds, there are no new attacks, vulnerabilities, or changes in risk specific to this topology that information security personnel need to consider. If public cloud services are chosen, changing security requirements will require changes to the network topology and the manner in which the existing network topology interacts with the cloud provider's network topology should be taken into account [7]. There are four significant risk factors in this use case:

- Ensuring the confidentiality and integrity of organization's data-in-transit to and from a public cloud provider;
- Ensuring proper access control (authentication, authorization, and auditing) to whatever resources are used at the public cloud provider;
- Ensuring the availability of the Internet-facing resources in a public cloud that are being used by an organization, or have been assigned to an organization by public cloud providers;
- Replacing the established model of network zones and tiers with domains.

4.2 Infrastructure Security – The Host Level

When reviewing host security and assessing risks, the context of cloud services delivery models (SaaS, PaaS, and IaaS) and deployment models public, private, and hybrid) should be considered [7]. The host security responsibilities in SaaS and PaaS services are transferred to the provider of cloud services. IaaS customers are primarily responsible for securing the hosts provisioned in the cloud (virtualization software security, customer guest OS or virtual server security).

4.3 Infrastructure Security – The Application Level

Application or software security should be a critical element of a security program. Most enterprises with information security programs have yet to institute an application security program to address this realm. Designing and implementing applications aims at deployment on a cloud platform will require existing application security programs to reevaluate current practices and standards. The application security spectrum ranges from standalone single-user applications to sophisticated multiuser e-commerce applications used by many users. The level is responsible for managing [7], [9], [10]:

- Application-level security threats;
- End user security;
- SaaS application security;
- PaaS application security;
- Customer-deployed application security
- IaaS application security
- Public cloud security limitations

It can be summarized that the issues of infrastructure security and cloud computing lie in the area of definition and provision of security specified aspects each party delivers.

5 Conclusion

The cloud is a major challenge in how computing resources will be utilized since aim of the cloud computing is to change the economics of the data center, but before sensitive and regulated data move into the public cloud, issues of security standards and compatibility must be addressed including strong authentication, delegated authorization, key management for encrypted data, data loss protections and regulatory reporting. All are elements of a secure identity, information and infrastructure model and can be applied to private and public clouds as well as to IAAS, PAAS and SAAS services. In the development of public and private clouds the service providers will need to use these guiding principles to adopt and extend security tools and secure products to build and offer end-to-end trustworthy cloud computing and services.

References

1. Cloud Computing, http://en.wikipedia.org/wiki/Cloud_computing
2. Cloud Computing: Business Benefits With Security, Governance and Assurance Perspectives. An ISACA Emerging Technology White Paper, http://www.isaca.org
3. Dhanjani, N., Rios, B., Hardi, B.: acking: The Next Generation. HO'Reilly Media, Inc., Sebastopol (2009)
4. ENISA.: Cloud Computing: Information Assurance Framework. ENISA (November 2009), http://www.enisa.europa.eu/
5. ENISA.: Cloud Computing: Benefits, risks and recommendations for information security. ENISA (November 2009), http://www.enisa.europa.eu/
6. Lyong, L.: How to Select a Cloud Computing Infrastructure Provider. Gartner, Inc. Research. ID Number: G00166565
7. Mather, T., Kumaraswamy, S., Latif, S.: Cloud Security and Privacy: An Enterprise Perspective on Risks and Compliance. O'Reilly Media, Inc., Sebastopol (2009)
8. Owens, K.: Securing Virtual Compute Infrastructure in the Cloud. White Paper: Cloud Computing, http://www.savvis.net
9. Reese, G.: Cloud Application Architectures: Building Applications and Infrastructure in the Cloud. O'Reilly Media, Inc., Sebastopol (2009)
10. Rittinghouse, J.W., Ransome, J.F.: Cloud Computing: Implementation, Management and Security. CRC Press, Boca Raton (2009)
11. Secure Cloud Architecture, NetApp, WP-7083-0809 (August 2009)
12. Security Guidance for Critical Areas of Focus in Cloud Computing V2.1, Cloud Security Alliance (2009), http://www.cloudsecurityalliance.org/guidance/csaguide.v2.1.pdf
13. Securing Microsoft's Cloud Infrastructure. A Microsoft White Paper, http://blogs.technet.com/gfs/archive/2009/05/27/securing-microsoft-s-cloud-infrastructure.aspx
14. The Role of Security in Trustworthy Cloud Computing. RSA, White paper, http://www.rsa.com

Security and Privacy Implications of Cloud Computing – Lost in the Cloud

Vassilka Tchifilionova

National Laboratory of Computer Virology,
Bulgarian Academy of Sciences
1113 Sofia, Bulgaria
vassie12@hotmail.com

Abstract. Cloud computing - the new paradigm, the future for IT consumer utility, the economy of scale approach, the illusion of un infinite resources availability, yet the debate over security and privacy issues is still undergoing and a common policy framework is missing. Research confirms that users are concern when presented with scenarios in which companies may put their data to uses of which they may not be aware. Therefore, privacy and security should be considered at every stage of a system design whereas advantages and disadvantages should be rated and compared to internal and external factors once a company or a person decides to go into the business of cloud computing or become just an user.

Keywords: Cloud computing, privacy, data, network, security, information.

> "As long as we live and breathe we'll be paranoid. We always have to be careful, but it isn't going to stop the movement of this technology" (David Barram)

1 Introduction

The need to search and meet ever increasing IT demands has often been seen and based on purely an economical base, transforming computing into a utility where performance and efficiency have become a part of the product. Internet was just the tool, the infrastructure that provided the platform for utilization of new services.

Cloud computing has already proved that it can reduce infrastructure costs and offer the ability to pay for services only when requested. Some experts even argue that services delivered via cloud computing should be used as a public utility [1].

J. Camenisch, V. Kisimov, and M. Dubovitskaya (Eds.): iNetSec 2010, LNCS 6555, pp. 149–158, 2011.

In a similar way, others argued that companies will increasingly purchase IT as a utility service from outside suppliers [2]. Others call it the "age of planetary computing" [3].

The cloud computing has been on the market for years (e.g. Microsoft Exchange/SharePoint, Google Apps, Amazon's Simple Storage Service, Twitter, Facebook and others) but it was analysed in details by professionals the last two years ago and it took some time until a common definition emerged. One of the very recent working definitions is brought up by the United States National Standards for Information Technology that defines cloud computing as a model for enabling convenient, on-demand network access to a shared pool of configurable resources that can be rapidly provisioned and released with minimal management effort or service provider interaction [4].

Cloud computing is based on abstraction of infrastructure, it is service oriented with the opportunity to offer dynamics and elasticity and at the same time minimise consumption and billing. The abstraction level takes part between the physical infrastructure and the owner of the information being stored and processed. The most common model of cloud computing consists of the three major components: software, platform and infrastructure.

Consumers are not paying for infrastructure, rather they pay for capability. For instance, Microsoft's product Exchange and SharePoint are available online for a monthly subscription instead of buying the full license for their use. This provides for a better utilization of computing resources. In addition, it has also an application in the national security domain which allows the intelligence community to execute missions at all levels by utilizing simultaneously the large volume of capabilities the technology can offer [5].

To understand the scale of cloud computing, one should look at analysis that estimate that within the next five years, the global market will grow to 95 billion dollars and that 12% of the worldwide software will move to the cloud [6]. Another study reveals that 66% of Americans connected to the Web use some kind of cloud service since most popular consumer-facing services that enable better collaboration are cloud based [7].

The most important question a company or a natural person should ask itself is not whether to move to a cloud but what part of the IT to move. Researchers argue that companies are forced to switch over to cloud computing in order to meet business' needs. Given the dynamic business environment and the focus on globalization, there are only a few enterprises that do not outsource some part of their business [8].

One can argue that this new approach is the future of networks capabilities and an innovative economical tool that would help companies to boost their performance by outsourcing fully their information systems and information technologies. But there are hidden costs and these are reputation, security and privacy. There are even experts that argue that the definition of cloud computing is unclear and that the benefits to the business are not being presented well [9]. The objective of this paper is to give the essence of security and privacy concerns that this new technological mean is facing and thus be a virtual checklist for all managers who consider or have considered the implementation of could computing approach. We are reaching a point where the implementation of cloud computing is inevitable.

2   Security Implications

It is well accepted in the business world that along with the benefits come risks and security concerns that must be taken into account. Managers should also consider the fact that new initiatives are most likely, unless fully examined, to bring potential for high risks. In general terms, information security is often associated with the tree principles of confidentiality, integrity and availability. Security risks such as viruses, Trojans, warms, spoofing, root kits and many others should not be ignored. Once they get into the cloud not only a single "customer" is at risk but the whole cloud with all its users. In that sense the security implication of a weak IDS and IPS will only increase the scale of damages caused not only to the cloud owner but to all the parties involved.

The very most well defined security concern is the abstraction between the physical infrastructure and the owner of the information being stored and processed. The traditional approach is based on the ownership of computer hardware where the data is stored. With cloud computing consumers do not own the computing infrastructure and often it is no longer clear who owns and controls the data if the third party is further outsourcing some of its infrastructure or services [10], [11] or if the real owner is not mentioned in the provider's term of services. For example, if a government agency owns the provider, terms of service that allows sharing with affiliates could result in all of the user's information being obtained by prosecutor or intelligence agencies without further notice or process. In most cases, the provider would also maintain a full record of user's activities and these records for instance might be well used in a law case [12].

There is always the risk of compromising confidential information by third parties. Users are required in most cases to establish their identity by providing personal information [13]. With regard to this, personal information could be misused if not properly protected. The cloud provider has also responsibility to safeguard this personal information in accordance with the data protection legislations.

In general, critics do not accept counter-arguments that cloud computing may reduce the number of severity of breaches in comparison to those suffered by desktop users, as well as users who store data with third parties outside the cloud [14].

Criticality of the data handled is yet another risk to be considered. Security is no longer of use if the data is not stored and managed in a proper way. The cloud is a very dynamic structure and in time of crises and disasters information can not be available not to exclude also the unavailability of Internet at the user's end because of his physical location. That is why an important part of any contract should be the incident response policy. It is unrealistic to expect that no incident will ever happen with a cloud service provider irrespective of the measures he has taken to restrict damages and impact. This can also be linked to the physical risk which is also a major security concern. The facility where the data is stored should have adequate measures for physical protection.

The level of risk can be determined by whether an organization is outsourcing a service or infrastructure or not. In that sense, in a multi-tier service provider arrangement, each of the parties involved share the risk of security.

Companies providing cloud services need to understand that authorisation is a key point when dealing with many customers. Methods such as two-factor authentication are desirable but it is a question of applicability. If a company uses this method it has to reflect on whether it is possible to transfer it to the cloud and what the implications would be?

Current cloud computing security is weakened by the fact that most companies don't have a cloud strategy yet [15] and those who have, very often, over time not update their security policies and it is quite possible that they do not meet the current threats and vulnerabilities which make the security procedure very "relaxed".

Users should be well aware of what they can and cannot do in the cloud and therefore it is critical to set up their rights and responsibilities at the very start. Yet another problem is again whether the cloud provider would support a role-based model which will allow users to know exactly their user rights and to access that particular data or service for which they are authorized to.

Many cloud providers use third-parties for security audit and so they show their willingness to be accountable and reliable to their customers. The audit itself could be followed by obtaining some form of an accreditation. In general there are two benefits of an audit report: it evaluates the security risks of the cloud provider and so it gives a feedback to the user and on the other hand, it is beneficial to the cloud provider for future improvements.

However, there are certain systems which are not subject to audit by external auditors since the legal regulations do not allow such. Such an example is the US Health Insurance Portability and accountability Act [16], [17].

There are different clouds with different security models and it is up to the manager to assess the services he would like to outsource or take advantage of and the residual risks that would be the result of his choice. The human factor should not be excluded [18]; it is indeed one of the highest security risks in any company. The people who are "running" the cloud should undergo a background check otherwise a company may face not only theft of data but also lack of competence in running a cloud service.

Once the expectation of the parties involved are clear, one can set up clear rules with regard to handling, use, storage and availability of information. A solution to the security problem would be the proper classification and labeling of information.

In addition to the above mentioned risk, there is also the possibility that the cloud provider may behave unfaithfully with regard to the users' outsourced data. For instance, for monetary reasons, reclaiming storage by discarding data that has been rarely accessed or not accessed at all, or in the worst case, hiding data loss as to maintain a reputation [16].

The common practice of cloud providers to encrypt data does not imply that clients should not encrypt data themselves before sending it [19]. However, if an intruder can figure out how to access your information in the cloud bypassing all authentication means he potentially can access all data and services.

Cloud services should be addressed clearly by information security policies of the companies.

A recent survey by ENISA indicates the most important risks: lock-in, failures in mechanisms separating customers' data and applications, and legal data protection legislation [20].

Many experts argue that unless one's company is in the security business it is quite clear that one's company would be less secure [13]. With that respect, cloud providers should use the principle of applying the highest security measures to the most risky client that would be in the cloud. Although security can be the weakest link of many cloud providers it can be also the strongest advantage of a service provider whose core business is information security. Therefore, one can argue that when we compare internal versus external information security we can be in favor of external.

Some cloud computing service providers argue that cloud computing is more secure than desktop-based and enterprise computing [3]. If the right security measures are in place on can argue that no data can ever be lost when a laptop is stolen or a desktop is attacked by unauthorized users since data is kept on private clouds and encrypted with access only provided to authorized users [2].

The reason for that is very simple: Many companies have IT security as their core business. In order to examine why IT security is hard to achieve we should consider the following [21]:

1. The number of highly skilled and experienced security technologists is very small.
2. Good security is expensive.
3. Your IT and security staff has an interest in the contents of your data.
4. Any resource (employees, contractors, service workers) in your organization that has access to your datacenter has access to all your data.
5. Most internal IT organizations grew with the business (Meaning that different systems over the time share common infrastructure).

And what happens if a vendor goes out of business?

To sum up, one of the few very important security issues before selecting a cloud vendor should cover privileged user access in order to know what people are oversighting such information and what the control over their access is. At second place, vendors should be willing to undergo an external audit and authentication. And last but not least, a vendor should state where exactly data is stored, under which jurisdiction and whether they will make a contractual commitment to obey privacy requirements [22].

3 Privacy Concerns

In the context of European data protection the issue of data security is very much bound to the data controller who is responsible and remains such for the collection and processing of personal data also in the cases where data is processed by a third party. Some European data protection authorities require data controllers to obtain a prior consent before data is transferred abroad and very often this is followed by a detailed description of why this needed and the means for protection of the receiving body [23].

It has to be stressed that not all types of could computing raise the same privacy and confidentiality risks. It is depends what type of information is published and the way it relates to a particular person and whether it was published with his consent. In reality these rights are ignored since consent has become a mechanism for guaranteeing continuous data flows, rather than a tool to protect individual rights [24].

The most common form of explicit consent nowadays is still the written contract. However, in the world of electronic transactions explicit consent is not that easy to come by. There is no comprehensive protective framework for safeguarding interests in the face of privacy of these new emerging technological applications.

Virtually every government worldwide would have a regulation that would allow an access to information that resides on a cloud. Cloud providers should discuss all these privacy and security implications and if necessary encrypt the data. This should be done in accordance to the current governmental legislation as not to be later considered a national threat.

A clear example for what the Government can do is the Patriot Act in the United States [25] where the federal government has the right to request details of ones online activities without the knowledge of the person. With that respect sometimes the forensics procedure may not be clear and so a third party can access your information in the cloud.

On the other hand, although the Government may have his legal right, a bigger concern should be the weak security systems of the websites where a phishing attack or key logging can happen easily. Even more than before, the importance of a strong password and any other form of authentication has came up so important [26].

It is quite interesting to note that information could be physically located in two or more different locations. And if the vendor moves this information without the knowledge of the owner then it is the case of changing the legislation and therefore suffering different legal consequences.

The trend toward adopting privacy and data protection legislation has continued consistently to the present. Most central European nations, as well as other jurisdictions, such as Australia, New Zealand and Hong Kong, now have so called "omnibus" data protection legislation.

The United States, however, has not adopted omnibus data protection legislation. The current culmination of three decades of European policy development is reflected in the European Directive 95/46/EC [27]on the Protection of

Individuals with the Regard to Processing of Personal Data and on the free Movement of such Data.

The EU Directive contains a significant extraterritorial provision that the flow of personal data from any EU member country may be halted, if the jurisdiction to which it is being transferred is deemed not to have an adequate level of protection for personal data. American companies doing business in Europe have had to adhere to the data protection laws of each of the jurisdictions in which they operate. As a matter of fact, many American companies know how to and have been complying with omnibus data protection legislation in the countries that require it. The irony is that these companies may not be providing the same level of protection of personal data for Americans or individuals in other parts of the world where there is no statutory requirement. Law seems to be powerless when it comes to international boundaries [28].

There is a common practice of cloud service providers to keep information in place after it has been removed from the user is in place. Most social networks misuse the information of their users and do not inform them about the privacy changes they made and in the worst case they inform them after the actual change leaving the users with no choice to delete or edit the information they have uploaded on the cloud.

In many instances a user may not be aware of the existence of a second-degree provider and often the cloud provider offers its facilities to the users without an individual concept. Changing the terms of the policy without limit would find a user a step back when he realizes that the privacy policy has changed and he has not had the opportunity to remove sensible data.

One of the greatest privacy concerns is the risk of compromise of confidential information which can pose a significant complication with regard to the fact that data could be handled and stored in different geographical locations. Compliance to regulations and laws in different geographical locations can be quite challenging. Although the legislation itself could be a solution to the privacy problem, during the years it has proved that in certain cases it would fail [23] Unless there is a legislation explicitly governing cloud computing, no security means would be able to substitute for it. It is often the case that law is behind the technology developments.

Another security risk posed by could computing is the mix up of information assets which is usually the result of high-availability systems without the set up of private networks [8].

In March 2009 the Electronic privacy information center filed a complaint and request for injunction before the Federal trade Commission exposing the security flaws of Google's Cloud Computing. In the above mentioned case, Google was unable to protect usernames and passwords allowing outsiders to "snoop on users' e-mail", another vulnerability was the security flow resulting from a vulnerability in Google desktop and Internet Explorer. [29].

Privacy is not only about companies' and enterprises' data but it is very often a matter of revealing personal information as it was in the case of Google. A more recent case was brought up to the light in February 2010 when the an

Italian judge decided to invoke a six-month suspended jail sentence on Google's global privacy council and two other company executives. This case proves the public need to maintain privacy and to make even executives accountable which is also believed to be the first time in privacy history [30].

Users should be enabled to give informed consent once the cloud provider proves that they are trustworthy and handle the information with care. The privacy commissioner of Ontario argues that companies need to understand that identity management is not only a business process but also a user activity. The infrastructure must account for many devices and allow for a unified user experience over all devices [31].

It is common for a cloud provider to offer cloud services or infrastructure without individual contracts. If the user is bound by the general published terms of services it would be very difficult to stop the provider to acquire a variety of rights on the information [12].

It could also happen that the cloud provider acquires information that reveals transactions and relations which may have purely a commercial value. The simplest example would two companies negotiating a merger or being rivals on the market.

The core of privacy that law protects should be clearly defined in terms of harmful uses and remedies. Imaginary harms must be addressed by communication and education, not by legislation and regulation [32], [33] Once people understand the business use of information, the benefits of free flow, and the cost of privacy, their privacy preferences may well appear to be not what we believe them to be. One of the latest developments in this area is the launch of the Privacy Impact Assessment by the UK Information Commissioners Office [34]. Similar Acts exist also in the USA, Canada and Australia where privacy commissioners are the primary privacy and data protection authorities. Most commonly, concerns arise when it is not clear why their personal information is requested or how it will be passed on to other parties [35].

4   Conclusions

Companies need to be very specific in choosing a provider. Accountability is one possible way to state the responsibility of the company offering cloud computing services independent of the place it resides or is processed. Still, legislation should be in place and must be a must.

There is no doubt that cloud computing is becoming more of a utility than simple capability delivered over a network. One big advantage is the concept of risk-share but if you do not know what you are sharing then you can not define and demand a proper security frame. Furthermore, only the minimum information should be collected or stores, and so inform the users what type of information one is storing and for what purposes.

Unless your organization in the security business it is likely that the company will be less secure than the cloud provider. For that reason, one should first compare the levels of security of his company and the cloud provider. Cloud computing has significant implications for the privacy and security of information not only for private but also for business information.

Security and privacy will remain a major concern until users become fully aware of the "depth" of the cloud: who manages it, how he does it and whether the company can afford to "give away" its information - decision that can only be taken after a careful risks analysis and policy considerations otherwise we may simply get lost in the cloud.

References

1. Mohamed, A.: A history of cloud computing (2009),
   http://www.computerweekly.com
2. Gourley, B.: Cloud Computing and Cyber Defense. A White Paper provided to the National Security Council and Homeland Security Council as input to the White House Review of Communications and Information Infrastructure, Crucial Point LLC (2009)
3. Reingold, B., Mrazik, R.: Cloud computing: the intersection of massive scalability, data security and privacy (Part I). Cyberspace Lawyer, [FNa1] 14 No. 5 GLCYLAW 1 Page 114 NO. 5, Cyberspace Law. 1, LegalWorks (2009)
4. United States National Standards for Information Technology,
   http://www.nist.gov/index.html
5. Dataline White Paper: Cloud computing for national security applications,
   http://www.dataline.com/soar.htm
6. Merrill Lynch, http://www.ml.com
7. Bruening, P., Treacy, B.: The Bureau of National Affairs, Inc. Cloud Computing: Privacy, Security Challenges, Privacy&Security Law. PVLR ISSN 1538-3423 (2009)
8. ISACA. Emerging Technology. White Paper Cloud Computing: Business Benefits With Security, Governance and Assurance Perspectives (2009)
9. Flinders, K.: Microsoft slashes cost of cloud computing (2009),
   http://www.computerweekly.com
10. Webbmedia Group's Knowledge Basem, Cloud Computing Explained (2009),
    http://www.webbmediagroup.com
11. Binning, D.: Top five cloud computing security issues (2009),
    http://www.computerweekly.com
12. Gellman, R.: Privacy in the Clouds: Risks to Privacy and Confidentiality from Cloud Computing, World Privacy Forum (2009)
13. Bertino, E., Paci, F., Ferrini, R., Shang, N.: Privacy-preserving Digital Identity Management for Cloud Computing, CS Department, Purdue University Bulletin of the IEEE Computer Society Technical Committee on Data Engineering
14. Reingold, B., Mrazik, R.: Cloud computing: Industry and government development (Part II)
15. Zimski, P.: Cloud computing faces security storm (2009),
    http://www.computerweekly.com
16. Wang, C., Wang, Q., Ren, K., Lou, W.: Privacy-Preserving Public Auditing for Data Storage: Security in Cloud Computing, Illinois Institute of Technology, Chicago IL 60616, USA, Worcester Polytechnic Institute, Worcester MA 01609, USA (2009)
17. 104th Congress, The Health Insurance Portability and Accountability Act (HIPAA) of, P.L.104-191 (1996),
    http://www.cms.hhs.gov/hipaageninfo/downloads/hipaalaw.pdf

18. Yee, A.: Cloud Computing Security Risks - Are they real (2009),
    http://www.ebizq.net
19. Andrei, T.: Cloud Computing Challenges and Related Security Issues, p. 5 (2009)
20. ENISA: Cloud Computing Benefits, risks and recommendations for information
    security, Report (2009)
21. Almond, C.: A Practical Guide to Cloud Computing Security: What you need to
    know now about your business and cloud security, Avande (2009),
    www.avande.com
22. Brodkin, J.: Gartner: Seven cloud-computing security risks, Network World (2009)
23. Tchifilionova, V.: The mirage of law - an oasis for surveillance. In: International
    Politics, vol. 4, pp. 7–23. South-west University "Neofit Rilski" Press, Blagoevgrad
    (2009), (in Bulgarian/English) ISSN 1312-5435
24. Davies, S.: Unprincipled privacy: why the foundations of data protection are failing
    us. University of New South Wales Law Journal (2001),
    http://www.austlii.edu.au/au/journals/UNSWLJ/2001/7.html
25. 107th Congress, The Patriot Act (2001),
    http://frwebgate.access.gpo.gov/cgi-bin/getdoc.cgi?dbname=107_cong_
    public_laws&docid=f:publ056.107.pdf
26. Trapani, G.: The Hidden Risks of Cloud Computing, Lifehacker (2009)
27. 95/46/EC Directive of the European Parliament and the Council (1995),
    http://ec.europa.eu/justice_home/fsj/privacy/docs/95-46-ce/
    dir1995-46_part1_en.pdf
28. Hurley, D.: A whole world in a glance: privacy as a key enabler of individual
    participation in democratic governance. Harvard Infrastructure Project, Harvard
    (2001), http://www.pco.org.hk/english/infocentre/conference.html
29. EPIC, Before the Federal Trade Commission, Complaint and request for Injunction.
    In the Matter of Google, Inc. and Cloud Computing Services
30. Vijayan, V.: Google privacy convictions in Italy spark outrage (2010),
    http://www.itworld.com
31. Cavoukian, A.: Privacy in the clouds: A White Paper on Privacy and Digital Iden-
    tity: Implications for the Internet, Information and privacy commissioner of On-
    tario (2009)
32. Bergkamp, L., et al.: EU Data Protection Policy. Computer Law and Security
    Report 18(1), 2002 (2002)
33. Litan, R.: Balancing Costs and Benefits of New Privacy Mandates. AEI-Brookings
    Joint Canter for Regulatory Studies, working paper 99-3 (April 1999)
34. The Information Commissioner's Office, PIA's Book (2007),
    http://www.ico.gov.uk/upload/documents/pia_handbook_html_v2/index.
    html
35. Pearson, S.: Taking Account of Privacy when Designing Cloud Computing Services.
    HP Laboratories, HPL-2009-54 (2009)

The Need for Interoperable Reputation Systems

Sandra Steinbrecher

Technische Universität Dresden, Fakultät Informatik, D-01062 Dresden, Germany
steinbrecher@acm.org

Abstract. Nowadays more and more Internet applications install reputation systems to collect opinions users have about some reputation objects. The opinions are usually formalized in the form of ratings the reputation system can use to build overall reputation profiles of the reputation objects. Reputation objects might be other users, products, web content and anything else that can be rated. Users may investigate the reputation object's reputation profile to estimate its quality resp. trustworthiness. As there are currently many providers of reputation systems it would be desirable to make reputation information in different systems interoperable or to establish meta reputation systems that collect information from various applications resp. their reputation systems. This process should consider both interoperability of reputation systems themselves and their interoperability with applications, trust and identity management systems as we will discuss in this paper.

1 Introduction

The experiences of an individual's environment influence his own interactions with others. When a friend made bad experiences with a shop the individual might become skeptical as well or even resist to buy there. The Internet offers its users the possibility to exchange knowledge and experiences regarding the online and offline world not only with others in their near environment, but with nearly everyone: Many users inform themselves about others' experiences with sellers before buying from them in a marketplace like eBay[1]. Many users inform themselves about hotels before booking a room in travel portals like tripadvisor[2]. Also many users make use of the book reviews collected in a store like amazon[3] before buying a book.

These three examples of eBay, tripadvisor and amazon have in common that the respective providers installed a reputation system that allows users to rate reputation objects of a certain category: in eBay other users, in tripadvisor objects related to traveling like hotels and restaurants, in amazon products like books the store sells.

Reputation systems can collect the experiences users make with reputation objects in a technically efficient way. These experiences may help other users to

---

[1] http://www.ebay.com/ (last visited April 2010)
[2] http://www.tripadvisor.com/ (last visited April 2010)
[3] http://www.amazon.com/ (last visited April 2010)

J. Camenisch, V. Kisimov, and M. Dubovitskaya (Eds.): iNetSec 2010, LNCS 6555, pp. 159–169, 2011.

estimate the future trustworthiness resp. quality of reputation objects they have no personal experience with. But informing himself about a reputation object does not prevent any user from making bad experiences with it because e.g., reputation usually is context-dependent and subjective. Although 'social attacks' (e.g., users may lie [6] or reputation objects may change) are possible, a usually large number of ratings and an honest majority of users will hopefully achieve that dissatisfied users are the exception. So reputation systems do not make other technical security measures (like digital signatures or certifications by independent institutions) obsolete, but hopefully reduce the cases where expensive legal enforceability might become necessary.[4]

Many users do not only use one, but several reputation systems, typically they might even be reputation objects in several reputation systems. E.g., both eBay and Amazon are providers of marketplaces and many people use both. For this reason there is an interest of users in using reputation they earned in one application as reputation object also in the other application. The same holds also for the author of a book: If he got good ratings on amazon, he might want to transfer these ratings also to the reputation system of another book store that sells his book. Also the respective store is interested in getting this reputation information from another store as it will typically increase its profit to provide large reputation profiles of the products it sells.

Reputation systems are now evolving into reputation-as-service applications like epinion[5] for products or iKarma[6] for companies/individuals independent from a concrete application, but still mostly have a single-provider model. There is the vision to establish stand-alone reputation systems that collect information from various interactions and in various contexts and also to make reputation information in different systems interoperable [9]. This process should consider both interoperability of reputation systems themselves and their interoperability with various other systems as we will discuss in this paper in Sect. 3-6 after explaining the preliminaries in Sect. 2.

2   Preliminaries

For our system environment as shown in Figure 1[7], we assume applications that allow users to make experiences with a so-called reputation object. Such an application might be, e. g., a marketplace where users make experiences with sellers or a wiki where users make experiences with content and authors. Further reputation systems are provided that collect positive and negative experiences the users report to it about reputation objects in the form of ratings given with a *rating function*. The reputation system updates the reputation of the reputation

---

[4] Please note that the social and legal aspects cannot be discussed in further detail.

[5] http://www.epinion.com (last visited April 2010)

[6] http://www.ikarma.com/ (last visited April 2010)

[7] Please note that we assume a user as reputation object in this Figure, but it can also be any arbitrary reputation object.

object from the ratings received with a *reputation function*. The reputation systems may exchange reputation between the reputation objects with the help of a *reputation exchange function*.

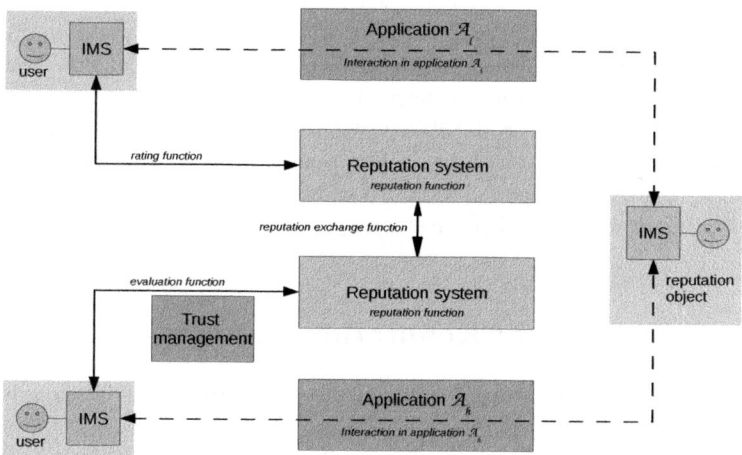

Fig. 1. System environment and arising interoperability issues

Reputation in a reputation system might be stored

- *centralized* at reputation servers designated for this purpose,
- *locally* with the reputation object itself, or
- *distributed* with other users.

If a user (the so-called evaluator) becomes interested in a reputation object, the reputation system provides him with an *evaluation function* to learn a reputation object's reputation following specific rules. The selection of ratings used by the evaluation function depends on both

- the *information flow* of ratings the system provides between the users and
- the *trust structure* between the users, i.e. how users trust in others' ratings.

The information flow is organized by the reputation system while the trust structure is determined by a trust management system that allows to assign trust values to other users. Depending on these two aspects the reputation selection for the evaluation function might be:

- *Global:* This means the information flow within the reputation network is complete and every evaluator gets the same reputation of a reputation object.
- *Individual:* This means an evaluator only gets a partial view on the reputation available. Here possibly every evaluator might receive a different reputation of the reputation object.

For interacting with any system users make use of an identity management system to possibly separate different partial identities they have.

From this architecture four interoperability issues arise that will be outlined in the following sections. First there is the aspect of interoperability between reputation systems that will be discussed in Sect. 3. This needs interoperability of the corresponding applications to either make reputation systems interoperable or to install a common reputation system that collects ratings from several applications as will be outlined in Sect. 4. In Sect. 5 we further discuss that for the evaluation function interoperability with trust management systems should become important. Finally in Sect. 6 the possibilities for interoperability with identity management systems for the reasons of privacy and security are explained. For further issues of reputation systems' privacy and security we refer to [18,9].

3   Interoperability of Reputation Systems

Typically users interact in manifold ways on the Internet. They might play, sell, post product ratings, discuss with friends on different topics and so on. As they might collect reputation in many applications, making existing reputation system(s) for these applications interoperable becomes of interest. The problem of interoperability that is represented by the reputation exchange function in our model is twofold:

- *Format:* First formats for common exchange and possibly also internal representation of reputation are needed. An OASIS group[8] works on a possible portable format using XML. But currently we still lack such a standard that could be implemented. Here is the need for solutions that can be easily integrated in the existing web technologies. The suggestion we implemented in [19] was to use the Resource Description Framework (RDF) [13] common for the Web 2.0 and allowing to add reputation information as meta-information to arbitrary web content.
- *Algorithm:* In every reputation system different implementations of rating function, evaluation function and reputation function are defined depending on the system designer. An overview of possible functions is for example given in [15]. For an economic introduction of possible advantages and disadvantages certain choices have we refer to [7]. An algorithm how to transfer reputations received from another reputation system to the own reputation system is needed. This algorithm needs to comprise inheritance rules for reputation to decide on interoperability of reputation or ratings from different reputation systems.

The OpenPrivacy Initiative[9] presented Sierra, a reference implementation of a reputation management framework comprising several components representing

---

[8] http://www.oasis-open.org/committees/tc_home.php?wg_abbrev=orms (last visited April 2010)
[9] http://openprivacy.org/ (last visited April 2010)

the functions of the reputation system as well as an identity management system. They also define reputation exchange functions, whose actual implementation can be determined by the system designer in terms of exchange rates between reputations calculated from different reputation functions.

However, there are other issues of interoperability between reputation systems so far neglected by the technical literature:

- *Several reputation exchanges:* For several executions of the reputation exchange functions between two reputation systems it has to be secured that it is clear which part of reputation has already been exchanged.
- *Related reputation objects:* So far we assumed that the reputation object is well determined. Another issue interoperability of reputation systems has to deal with is the possible relation between distinct reputation objects. For example a reputation system collecting reputation of content might need to exchange reputation with a reputation system collecting reputation of authors. Certainly there is some relation between a content and its authors, but it might not be advisable to transfer reputation of one content directly to its authors and vice versa. Thus reputation systems need to define the transfer of reputation between related objects by a *reputation object exchange function.*

4   Interoperability with Applications

Currently the vision arises to establish stand-alone reputation systems that collect information from various interactions in different applications.

Social scientists and theoretical economists model the problem whether two users, who want to interact, should place trust in each other as a so-called trust game [2,5] that needs inter-personal context-specific trust.

The reputation system tries to assist users in this game by implementing a social network that allows users to exchange information with each other. By the evaluation function users can learn others' reputation. In social sciences this is called the **learning mechanism** of the social network [1]. On the other hand users may control other users by spreading information about the users in the social network. In social sciences this is called the **control mechanism** of the social network [1] as implemented with the rating function of the reputation system. Thus, he applications, where the interactions rated took place, have to provide the reputation system with as much information as possible on the following aspects:

- *Model of Trust Game:* Only users, who gave a leap of faith to reputation objects should be able to rate them. Applications have to make a clear model, who gave a leap of faith and specify this for the reputation system.
- *Interaction information:* As reputation is context-dependent information on the interaction rated is needed, e.g., time, value for the interaction partners.
- *Rater information:* As reputation needs to build on inter-personal trust also information on the raters is needed as will be outlined in Sect. 5

Beneath the OpenPrivacy Initiative mentioned in Sect. 3 there are commercial stand-alone systems like iKarma as 'third-party service for collecting, managing and promoting [your] reputation among [your] customers and contacts.'[10] or portals like Trivago[11] that comprises reputation information from various other reputation systems.

The scientific approaches, that outline reputation infrastructures independent from concrete applications (e.g., [20,16,11,17], do not follow the centralised approach of the commercial solutions, but use local storage of reputation information to enable users to show the reputation they collected to others themselves. All of these suggestions need some external infrastructure to prevent reputation manipulation by the reputation object.

In the mentioned scientific approaches the trust model is implicitly clear, but as all of them aim for a privacy-respecting reputation system neither interaction nor rater information is provided. For the commercial solutions users can provide as much information as they want on themselves and their interactions.

5   Interoperability with Trust Management

As outlined in Sect. 2 reputation networks need to have some kind of inherent trust structure. When a user wants to determine a reputation object's credibility resp. trustworthiness he has to determine his trust in two other sources as well:

- *Raters:* The ratings given by raters can be:
  - *subjective ratings*, that are influenced by the raters' subjective estimation of the reputation object, or
  - *objective ratings*, that can be verified by all other users than the rater at some point in time and that would have come to the same ratings.

  An example for the first type of ratings is eBay while examples for the second type can be found in P2P systems, e.g. GNUnet[12], where the reply to a query leads to a positive reputation, and a reply can be proved or verified at least at the time it is sent.

  If the raters are humans as in our model from Sect. 2, subjective ratings will be given. Then the rater needs to decide whether he would have come to the same rating; this means their views on the reputation object is interoperable. For this reason a trust management system to determine the inter-personal trust in raters is needed. It can be realized by an additional reputation system for raters.
- *Reputation systems:* Evaluators need to have system trust in all reputation systems that collected the ratings and calculated the reputation the user evaluates.

Technically trust management is often associated with PKI structures [14] (beneath other approaches). PKI structures allow to bind keys to pseudonyms.

---

[10] http://ikarma.com/support/faq/#1 (last visited April 2010)
[11] http://www.trivago.com/ (last visited April 2010)
[12] www.gnunet.org (last visited April 2010)

Others can use their key to sign this binding. Thereby chains to other users, who want to trust in this binding can be built. These chains can be done hierarchically with certification authorities or in the form of the web of trust (e.g., GPG/PGP). Both structures could and should also be used for the broader deployment of reputation systems. Hierarchies and chains as they work for trust management could be applied to reputation management to express which experiences from others can be trusted.

However, the straightforward approach to implement ratings as signatures and use existing PKI structures only assures accountability of keys and linkage to their holder. But if a user or certification authority signs someone's key in a PKI structure that does not say anything about the credibility/ competence they assume the key holders to have as reputation objects. For this reason different key(s) than for accountability are needed and eexisting certificate structures have to be extended appropriately.

6   Interoperability with Identity Management

For the evaluation function of reputation systems not only the overall reputation, but also the single ratings and the raters, who gave them might be important. If raters misbehave maliciously by giving ratings, that do not reflect the concrete experience they made with reputation objects, there should be a possibility to detect this and probably to make them accountable for that.

But for the collection of large reputation profiles about users (both reputation objects and raters) privacy also becomes an important issue. Reputation systems often collect information about who interacted with whom in which context. Such information should be protected by means of technical data protection to ensure users' right of informational self-determination [12].

For managing this a reputation system should be interoperable with privacy-enhancing user-controlled identity management systems (PE-IMS). An IMS in general is able to certify users and grant rights to them for applications. Additionally a PE-IMS [3,4] like PRIME[13] assist users platform-independent in controlling their personal data in various applications and selecting pseudonyms appropriately depending on their wish for pseudonymity and unlinkability of actions.

The interoperability of a reputation system with a PE-IMS needs a privacy-respecting design of reputation systems while keeping the level of trust provided by the use of reputations as outlined in [18].

When a reputation system interoperates with a PE-IMS it is possible and intended that users have several partial identities (pIDs) which cannot be linked, neither by other users using the systems nor by the underlying system (as long as the user does not permit this). Both raters and reputation objects might only be known by pseudonyms to each other.

---

[13] Privacy and Identity Management for Europe (http://www.prime-project.eu/), funded by the European Union in the 6. Framework Program, 2004-2008.

If there would exist only one reputation per user, all pIDs of this user would have the same reputation. This would ease the linking of the pIDs of one user because of the same reputation value. Thus, having separated reputations per pID and not only one per user is a fundamental condition for a reputation system in the context of identity management.

The use of pIDs arises the problem that a malicious user may rate himself a lot of times using new self created pID for every rating in order to improve his own reputation. This kind of attack is also known as Sybil attack [8]. If the reputation system is not defined carefully, it would be easy for such an attacker to improve the own reputation unwarranted. This can be limited/prevented by entrance fees or the use of once-in-a-lifetime credentials as suggested in [10]. When using PRIME as IMS the latter can be implemented by its identity provider issuing such credentials. Alternatively or additionally also fees could be collected.

7   Resulting Infrastructure

For users as reputation objects we outline in the following a possibly resulting secure reputation system interoperable with an application, an identity and trust management. Our design description is independent from concrete rating, reputation and evaluation functions.

We assume all communication to be secured by encryption to reach confidentiality of all ratings and actions performed. Also all messages should be transferred in an anonymous way with an anonymous communication network. All actions and ratings have to be secured by digital signatures (given under a pseudonym) for integrity reasons.

For the identity management a user registers himself with an identity management system (provider) by declaration of his identity data (step 1 in Fig. 2). After verifying the data the identity provider issues a credential or certification on (part of) these data (step 2 in Fig. 2). By the use of an identity management system (provider) accountability of the pseudonym can be given.

When the user wants to register with a reputation system (provider) he sends it the certification/credential he got from the identity management system (provider) (step 3 in Fig. 2). This should guarantee that no user is able to build up reputation under multiple pseudonyms within the same context and every user can be identified in the case of misbehavior. The reputation system (provider) creates a reputation certificate/credential based on the certificate/-credential from the identity management system (provider) and sends it back to the user (step 4 in Fig. 2).

The reputation credential contains the user's reputation pseudonym, its initial reputation and possibly other attributes like the applications it can be used in or an expiration date.

Based on the reputation credential the user can register himself with an application by showing his reputation certificate/credential (step 5 in Fig. 2). Thereby

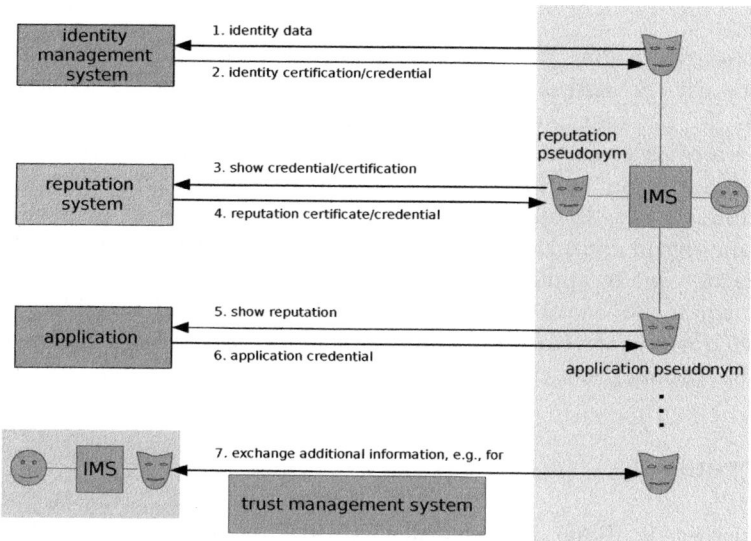

Fig. 2. Infrastructure for users as reputation objects

he agrees that he will collect reputation for his interactions within the application (e.g., a marketplace or a wiki) with the reputation system he registered with. Based on this he gets an application credential to use the application (step 6 in Fig. 2).

Additionally the user might interact with other users to exchange additional information, e.g. via a trust management system to inform himself about this user (possibly as a rater) and other users in the reputation network (step 7 in Fig. 2).

Every action the user performs above can be done under distinct pseudonyms if convertible credentials are issued by the respective providers. We implemented this infrastructure for phpBB as application and the user-controlled privacy-enhancing identity management PRIME as outlined in [16]. Currently we lack a trust management in our implementation.

8   Conclusion

In this paper we gave a first impression which aspects of interoperability should be considered for reputation systems. We also described a possible infrastructure for interoperability between applications, reputation, trust and identity management systems from a technical perspective. For interoperability of reputation systems themselves and implementing corresponding rating, reputation and evaluation functions an overall treatment from various scientific disciplines will be needed to come to suitable solutions usable in practice.

Acknowledgements

The research leading to these results has received funding from the European Communitys Seventh Framework Programme (FP7/2007-2013) for the project PrimeLife. The information in this document is provided as is, and no guarantee or warranty is given that the information is fit for any particular purpose. The PrimeLife consortium members shall have no liability for damages of any kind including without limitation direct, special, indirect, or consequential damages that may result from the use of these materials subject to any liability which is mandatory due to applicable law.

For comments on preliminary versions of this papers I would like to thank the anonymous reviewers. Additionally Stephan Groß and Vashek Matyas provided valuable comments.

References

1. Buskens, V., Raub, W.: Embedded trust: Control and learning. In: Lawler, E., Thye, S. (eds.) Group Cohesion, Trust, and Solidarity. Advances in Group Processes, vol. 19, pp. 167–202 (2001)
2. Camerer, C., Weigelt, K.: Experimental tests of a sequential equilibrium reputation model. Econometrica 56, 1–36 (1988)
3. Clauß, S., Pfitzmann, A., Hansen, M.: E Van Herreweghen. Privacy-enhancing identity management. The IPTS Report 67, 8–16 (2002)
4. Clauß, S., Köhntopp, M.: Identity management and its support of multilateral security. Computer Networks 37(2), 205–219 (2001)
5. Dasgupta, P.: Trust as a commodity. In: Gambetta, D. (ed.) Trust: Making and Breaking Cooperative Relations. Department of Sociology, pp. 49–72. University Oxford, Oxford (2000)
6. Dellarocas, C.: Immunizing online reputation reporting systems against unfair ratings and discriminatory behavior. In: EC 2000: Proceedings of the 2nd ACM Conference on Electronic Commerce, pp. 150–157. ACM Press, New York (2000)
7. Dellarocas, C.: The digitization of word-of-mouth: Promise and challenges of online feedback mechanisms. Management Science, 1407–1424 (October 2003)
8. Douceur, J.R.: The sybil attack. In: Druschel, P., Kaashoek, M.F., Rowstron, A. (eds.) IPTPS 2002. LNCS, vol. 2429, pp. 251–260. Springer, Heidelberg (2002)
9. ENISA. Position paper. reputation-based systems: a security analysis (2007), http://www.enisa.europa.eu/doc/pdf/deliverables/enisa_pp_reputation_based_system.pdf (letzter Abruf September 02, 2008)
10. Friedman, E., Resnick, P.: The social cost of cheap pseudonyms. Journal of Economics and Management Strategy 10, 173–199 (1999)
11. Kumar, S.S., Koster, P.: Portable reputation: Proving ownership across portals. In: Proc. of the European Context Awareness and Trust 2009 (EuroCAT 2009), 3rd Workshop on Combining Context with Trust, Security, and Privacy. CEUR Workshop Proceedings, vol. 504, pp. 21–30 (September 2009)
12. Mahler, T., Olsen, T.: Reputation systems and data protection law. In: eAdoption and the Knowledge Economy: Issues, Applications, Case Studies, pp. 180–187. IOS Press, Amsterdam (2004)

13. Manola, F., Miller, E.: RDF Primer. W3C Recommendation, W3C (February 2004), http://www.w3.org/TR/rdf-primer/ (last visited July 01, 2009)
14. Maurer, U.: Modelling a public-key infrastructure. In: Bertino, E. (ed.) ESORICS 1996. LNCS, vol. 1146, pp. 325–350. Springer, Heidelberg (1996)
15. Mui, L.: Computational Models of Trust and Reputation: Agents, Evolutionary Games, and Social Networks. PhD Thesis, Massachusetts Institute of Technology (2003)
16. Pingel, F., Steinbrecher, S.: Multilateral secure cross-community reputation systems. In: Furnell, S.M., Katsikas, S.K., Lioy, A. (eds.) TrustBus 2008. LNCS, vol. 5185, pp. 69–78. Springer, Heidelberg (2008)
17. Schiffner, S., Clauß, S., Steinbrecher, S.: Privacy and liveliness for reputation systems. In: Martinelli, F., Preneel, B. (eds.) EuroPKI 2009. LNCS, vol. 6391, pp. 209–224. Springer, Heidelberg (to appear, 2010)
18. Steinbrecher, S.: Enhancing multilateral security in and by reputation systems. In: Matyáš, V., Fischer-Hübner, S., Cvrček, D., Švenda, P. (eds.) IFIP WG 9.2, 9.6/11.6, 11.7/FIDIS. IFIP Advances in Information and Communication Technology, vol. 298, pp. 135–150. Springer, Heidelberg (2009)
19. Steinbrecher, S., Groß, S., Meichau, M.: Jason: A scalable reputation system for the semantic web. In: Gritzalis, D., Lopez, J. (eds.) SEC 2009. IFIP AICT, vol. 297, pp. 421–431. Springer, Heidelberg (2009)
20. Voss, M.: Privacy preserving online reputation systems. In: International Information Security Workshops, pp. 245–260. Kluwer, Dordrecht (2004)

Author Index

Almgren, Magnus   29
Amran, Ahmad R.   134
Arnab, Alapan   56

Cavedon, Ludovico   88
Conti, Mauro   20
Crispo, Bruno   20

Dimitrov, Kiril   29
Di Pietro, Roberto   20
Djambazova, Edita   29

Hutchison, Andrew   56

Ivanov, Veliko   104

Jonsson, Erland   29
Jordaan, Louis   70

Kellermann, Benjamin   9
Kisimov, Valentin   104
Kruegel, Christopher   88

Martin, Tobias   56
Murdjeva, Alexandra   104

Nickolov, Eugene   123

Ortolani, Stefano   20

Palazov, Anton   117
Parish, David J.   47, 134
Phan, Raphael C.-W.   47, 134
Polimirova, Dimitrina   123

Steinbrecher, Sandra   159

Tchifilionova, Vassilka   149
Tzaneva, Monika   104

Velev, Dimiter   140
Vigna, Giovanni   88
von Solms, Basie   1, 70

Whitley, John N.   47, 134

Zlateva, Plamena   140

GPSR Compliance

The European Union's (EU) General Product Safety Regulation (GPSR)
is a set of rules that requires consumer products to be safe and our
obligations to ensure this.

If you have any concerns about our products, you can contact us on
ProductSafety@springernature.com

In case Publisher is established outside the EU, the EU authorized
representative is:

Springer Nature Customer Service Center GmbH
Europaplatz 3
69115 Heidelberg, Germany

Batch number: 09478804

Printed by Printforce, the Netherlands